Sharing Network Resources

Critical Acclaim for *Sharing Network Resources*

"This book is beautifully written and will be a major resource for graduate students and researchers. I plan to have my graduate students read it, both for its intellectual content and the elegant presentation."

Professor R. Srikant
Fredric G. and Elizabeth H. Nearing Endowed Professor,
University of Illinois at Urbana-Champaign

"This is a very nice book. I look forward to seeing it published and using it in my future offerings of courses on network economics."

Professor Jianwei Huang
The Chinese University of Hong Kong

"I commend the authors for having successfully brought a diverse set of analytical tools developed in last 30 years from research communities such as economics and operations research into a networking textbook. I am particularly impressed that the authors are able to make these cutting-edge tools accessible to engineering students with little sacrifice for their mathematical rigor and generality."

Jim Dai
Professor of Operations Research and Information Engineering, Cornell University
The Chandler Family Chair of Industrial and Systems Engineering,
Georgia Institute of Technology (on leave)

"This is indeed a fascinating book, containing many nice results (both classical and new). It covers several key aspects of sharing network resources that are not usually covered in a typical networking book. Mostly based on mathematical models, the book deals with principles instead of protocol details."

Libin Jiang
Qualcomm Research

Synthesis Lectures on Communication Networks

Editor
Jean Walrand, *University of California, Berkeley*

Synthesis Lectures on Communication Networks is an ongoing series of 50- to 200-page publications on topics on the design, implementation, and management of communication networks. Each lecture is a self-contained presentation of one topic by a leading expert or team of experts. The topics range from algorithms to hardware implementations and cover a broad spectrum of issues from security to multiple-access protocols. The series addresses technologies from sensor networks to reconfigurable optical networks. The series is designed to:

- Provide the best available presentations of important aspects of communication networks.
- Help engineers and advanced students keep up with recent developments in a rapidly evolving technology.
- Facilitate the development of courses in this field

Sharing Network Resources
Abhay Parekh and Jean Walrand
2014

Wireless Network Pricing
Jianwei Huang and Lin Gao
2013

Performance Modeling, Stochastic Networks, and Statistical Multiplexing, second edition
Ravi R. Mazumdar
2013

Packets with Deadlines: A Framework for Real-TimeWireless Networks
I-Hong Hou and P. R. Kumar
2013

Energy-Efficient Scheduling under Delay Constraints forWireless Networks
Randall Berry, Eytan Modiano, and Murtaza Zafer
2012

NS Simulator for Beginners
Eitan Altman and Tania Jiménez
2012

Network Games: Theory, Models, and Dynamics
Ishai Menache and Asuman Ozdaglar
2011

An Introduction to Models of Online Peer-to-Peer Social Networking
George Kesidis
2010

Stochastic Network Optimization with Application to Communication and Queueing Systems
Michael J. Neely
2010

Scheduling and Congestion Control forWireless and Processing Networks
Libin Jiang and Jean Walrand
2010

Performance Modeling of Communication Networks with Markov Chains
Jeonghoon Mo
2010

Communication Networks: A Concise Introduction
Jean Walrand and Shyam Parekh
2010

Path Problems in Networks
John S. Baras and George Theodorakopoulos
2010

Performance Modeling, Loss Networks, and Statistical Multiplexing
Ravi R. Mazumdar
2009

Network Simulation
Richard M. Fujimoto, Kalyan S. Perumalla, and George F. Riley
2006

Sharing Network Resources

Abhay Parekh and Jean Walrand

ISBN: 978-3-031-79265-6 print
ISBN: 978-3-031-79266-3 ebook

DOI: 10.1007/978-3-031-79266-3

A Publication in the Springer series
SYNTHESIS LECTURES ON COMMUNICATION NETWORKS
Series ISSN: 1935-4185 print 1935-4193 ebook

Lecture #15
Series Editor: Jean Walrand, *University of California, Berkeley*

First Edition

10 9 8 7 6 5 4 3 2 1

Sharing Network Resources

Abhay Parekh
University of California, Berkeley

Jean Walrand
University of California, Berkeley

SYNTHESIS LECTURES ON COMMUNICATION NETWORKS #15

ABSTRACT

Resource Allocation lies at the heart of network control. In the early days of the Internet the scarcest resource was bandwidth, but as the network has evolved to become an essential utility in the lives of billions, the nature of the resource allocation problem has changed. This book attempts to describe the facets of resource allocation that are most relevant to modern networks. It is targeted at graduate students and researchers who have an introductory background in networking and who desire to internalize core concepts before designing new protocols and applications.

We start from the fundamental question: what problem does network resource allocation solve? This leads us, in Chapter 1, to examine what it means to satisfy a set of user applications that have different requirements of the network, and to problems in Social Choice Theory. We find that while capturing these preferences in terms of utility is clean and rigorous, there are significant limitations to this choice. Chapter 2 focuses on sharing divisible resources such as links and spectrum. Both of these resources are somewhat atypical—a link is most accurately modeled as a queue in our context, but this leads to the analytical intractability of queueing theory, and spectrum allocation methods involve dealing with interference, a poorly understood phenomenon. Chapters 3 and 4 are introductions to two allocation workhorses: auctions and matching. In these chapters we allow the users to game the system (i.e., to be strategic), but don't allow them to collude. In Chapter 5, we relax this restriction and focus on collaboration. Finally, in Chapter 6, we discuss the theoretical yet fundamental issue of stability. Here, our contribution is mostly on making a mathematically abstruse subdiscipline more accessible without losing too much generality.

KEYWORDS

auctions, economics, matching, networks, optimization, resource allocation, stability, utility

Dedicated to Kadambari and Annie

Contents

List of Figures

List of Tables

Preface

The idea to write this book came to us while teaching a graduate networking course in the Spring of 2009 at Berkeley. We asked our students to read a number of recent papers in the field, and quickly realized that while they understood the mechanics of networks (for example, how packets are routed), they had a hard time relating analytical tools from other fields (such as Game Theory and Operations Research) to the study of communication networks. We started out to write a text that summarized known results, but this turned out to be unwieldy and duplicative.

The book we ended up writing cherry-picks from the vast literature to focus attention on aspects of resource allocation that we think are especially relevant to modern networks. In some cases, this led us to eliminate standard topics; for example, we dispensed with Queuing Theory except to talk about stability. On the other hand, we included areas that are usually glossed over. For example, the growing diversity of applications delivered over networks led us to include a treatment of Social Choice Theory. We also included more "human" activities such as collaboration and auctions since networks are increasingly becoming platforms for such activities. The end result is a short book with references to the literature. Every topic that we cover reflects our biases, and attempts to provide our take on why it is important.

Our audience is the mathematically curious reader who is somewhat familiar with Networking (at the level of [Kurose et al. 2003, Bertsekas and Gallager 1987, Peterson and Davie 2007, Walrand and Parekh 2010]). We assume some background in probability, calculus, and algorithms, but nothing beyond preparation at the undergraduate level. The mathematical treatment allows us to be more conceptual and to bring basic, yet beautiful, ideas from social choice theory, random processes, and convex optimization to the realm of networking. However, such an approach suffers from the distinct possibility of becoming too theoretical and of no real use to an engineer who wants to actually build network devices and protocols. In order to prevent us from falling into that trap, we have kept the mathematics accessible and, as mentioned earlier, we have eliminated a number of topics for which connecting the dots from theory to practice seemed a bit too tenuous. Our hope is that readers looking to actually build network protocols and applications will be inspired to take the ideas we discuss and make them practical.

ORGANIZATION OF THE BOOK

In contrast to most introductory books, ours is not organized by the layers of the protocol stack. Many networking innovations come from thinking across these layers or in between them, and so our focus is on general principles as opposed to protocol details. Our book should be read after the reader is comfortable, for example, with the function of flow control and with the details of TCP.

We start from the fundamental question: What problem does network resource allocation solve? This leads us to examine what it means to satisfy a set of applications (users) that have different requirements of the network, and to problems in Social Choice Theory. We find that while capturing these preferences in terms of utility is clean and rigorous, there are significant limitations to this choice. Chapter 2 focuses on sharing divisible resources such as links and spectrum. Both of these resources are somewhat atypical—a link is most accurately modeled as a queue in our context, but this leads to the analytical intractability of queueing theory, and spectrum allocation methods involve dealing with interference, a poorly understood phenomenon. Chapters 3 and 4 are introductions to two allocation workhorses: auctions and matching. In these chapters we allow the users to game the system (i.e., to be strategic), but don't allow them to collude. In Chapter 5 on collaboration, we relax this restriction. Finally, in Chapter 6, we discuss the theoretical yet fundamental issue of stability. Here, our contribution is mostly on making a mathematically abstruse sub-discipline more accessible without losing too much generality.

ACKNOWLEDGMENTS

A number of students and colleagues read drafts of this book and helped make it more readable. We would like to thank Jim Dai (Georgia Tech), Lingjie Duan (SUTD), Jianwei Huang (CUHK), Longbo Huang (Tsinghua), Libin Jiang (Qualcomm), John Musacchio (UCSC), Ramesh Johari (Stanford), R. Srikant (UIUC), Stephan Adams, Vijay Kamble, Ramtin Pedarsani, Galina Schwartz, Nihar Shah (UCB), and the students in EE228. Thanks to Vivek Borkar (IIT Bombai) who was kind enough to host Abhay Parekh at the Tata Institute of Fundamental Research in Mumbai while he worked on an early version.

CHAPTER 1
Social Choice

The resources of a network (its routers, links, and servers) are designed to be shared among many users. When demand exceeds supply, network protocols and mechanisms implement policies or rules to decide how to trade off the satisfaction of users. There are two separable issues: What are the best policies, and what are best mechanisms to implement a given set of policies? This chapter focuses on the first issue, and deals with questions such as: Should a router give priority to voice over video? Should a data center allocate its resources on a first come-first served basis?

However, the quandaries presented by these questions are shared by even more basics: How should one share a pie? Should a hungrier person get a larger share or should it be divided it equally? How should a couple with different preferences decide whether to go to the opera or play tennis?

Such basic questions suggest that before even thinking of designing an algorithm to share network resources, we should reflect on the fundamental difficulties of resource allocation.

The problem of allocating resources to attempt to meet the desires of a collection of users is called a *social choice problem*. The goal of this chapter is to expose the reader to some of the issues faced in formulating and solving such a problem. We begin with formulations in which options are ranked by preference. We show through simple examples that ordinal preferences generally do not have to be transitive, and even when they are, it is usually not possible to rank the options in a consistent fashion that "makes sense." Next, we formulate social choice problems in the context of utility, which is a measure of an option's desirability. While utilities are more expressive than preferences, they have significant drawbacks as well. The point of these somewhat negative results is to make the reader aware of the tradeoffs involved in choosing a formulation rather than to argue against moving forward with confidence.

1.1 PREFERENCES

When comparing options, it is natural for individuals to have preferences. Thus, you may prefer coffee over tea and chocolate ice cream over strawberry sherbet. We discuss possible characteristics of such preferences and how to combine the preferences of different people.

One of the first properties that one might require is that preferences be *transitive*. What this means is that if A is preferred to B and B to C, then it must be that A is preferred to C. This allows us to assign numerical values to the options that reflect preferences. But as the next three examples show, transitivity sometimes doesn't hold even in natural settings. This makes the task of allocating resources challenging. As will become evident, problems arise when preference is a function of more than one attribute. Even though an individual may have transitive preferences for each attribute for equal values of the other attributes, this may not hold in general.

Example 1.1 (Choosing a Car) Bob tries to choose a car. He prefers a cheaper car, unless the difference in comfort is substantial. He has narrowed down the final selection to four models: Corolla, Camry, Lexus 250sc, and Mercedes E350. Ideally, Bob could assign a numerical value to each reflecting how much he values each car and pick the car with the highest value. But given his preferences, he prefers a Corolla to a Camry, a Camry to the Lexus, and the Lexus to the Mercedes. However, given the choice between a Corolla and the Mercedes, he prefers the Mercedes since it is far more comfortable. For any of the cars, *there is always another one that he prefers*, and so it isn't possible to map Bob's degree of preference into a numerical value.

Example 1.2 (Low Delay or High Throughput?) Different applications have various preferences in terms of delay and throughput. Smaller delay and larger throughput are preferable, but how can one trade off delay and throughput? For instance, a video connection with a lower resolution may experience a smaller delay. Is that preferable?

It is conceivable that a larger throughput is preferable, unless the delay becomes significantly larger. We denote the delay and throughput by a pair (d, r) where d is a delay in ms and r a rate in kbps. By $(d_1, r_1) \succeq (d_2, r_2)$, one indicates that one prefers the connections with the characteristics (d_1, r_1) to the other. One might have the following preferences:

$$(110, 140) \succeq (95, 100), (95, 100) \succeq (85, 80), (85, 80) \succeq (80, 70)$$

but

$$(80, 70) \succeq (110, 140).$$

Our final example shows that non-transitivity lurks even in cases where our intuition tells us it should not.

Example 1.3 (The Winning Die) Bob has to pick one of three six-sided dice:

- die A has sides: 2, 2, 4, 4, 9, 9
- die B has sides: 1, 1, 6, 6, 8, 8
- die C has sides: 3, 3, 5, 5, 7, 7

After Bob picks his die, Alice gets to pick one of the other two. Each player rolls their chosen die and the highest roll wins. Which die should Bob pick? Bob's preferences can be expressed as follows: Given two dice he prefers the one that has greater than even odds of rolling higher than the other. Then, he must prefer A over B, B over C, but C over A! [1]

For more examples of intransitivity see [Fishburn 1991].

TENNIS ANYONE? ARROW'S IMPOSSIBILITY THEOREM

Since intransitive preferences greatly complicate the problem of assigning resources, let us assume that preferences are transitive. Suppose Alice has the following preferences:

$$movie \succeq opera \succeq tennis$$

and Bob's preferences are

$$tennis \succeq movie \succeq opera.$$

Alice and Bob want to share an activity. Which one should they pick?

A simple way to make the decision is to always follow Bob's preference, in which case we say that Bob is a dictator, but that is not fair. Another way is to rule out certain alternatives that they both dislike. For example, they would both rather watch a movie than go to the opera, so perhaps we could just rule out the opera.

More generally, it would be nice to have a process by which we could take as inputs Alice and Bob's preferences and output a single alternative (such as movie) that is consistent with certain common-sense rules of fairness. Such a process is called a *Social Choice Function (SCF)*. Let us call the arguments (the individual preference functions) to the SCF a *Profile*. Here are some reasonable properties of an SCF:

1. *No dictator.*
2. *Unanimity*: If $A \succeq B$ in every individual order, then B can't be picked as the best choice.
3. *Monotonicity*: Suppose X is picked over Y for profile R. Then, X is also picked over Y for every other profile R' such that the preferences of Alice and Bob for X over Y are the same or stronger than in R.

Remarkably, Arrow [Arrow 1950] showed that there is no SCF that meets these properties when there are at least three items. So much for marriage counseling!

---

1. Let p_{ij} be the probability that die i beats die j. Then summing over the possible outcomes: $p_{AB} = P_{BC} = \frac{5}{9}$ but $p_{AC} = 1 - p_{CA} = \frac{4}{9}$

We will not prove the theorem (see [Mas-Colell et al. 1995]), but let's see why this is the case for our example. To simplify notation, we designate the three activities by T, O, M and the social choice function as f that takes a profile as its argument. Write TMO for $T \succeq M \succeq O$, and similarly for the permutations. Thus, $f(TMO, OTM)$ is the choice in $\{T, O, M\}$ when the preferences of Alice are TMO and those of Bob are OTM.

If $f(TOM, MOT) = T$, then Alice is a dictator. To see this, we claim that $f(TOM, XYZ) = T$ for any permutation XYZ of TOM. If $X = T$ we are done from the first condition. For any other permutation, condition (3) forces f to map to T. Similarly, if $f(TOM, MOT) = M$, then Bob is a dictator.

Thus, we must assume that $f(TOM, MOT) = O$. Clearly, $f(TOM, MTO) \neq O$ from (2). Also, if $f(TOM, MTO) = M$ then $f(TOM, MOT) = M$ which contradicts the assumption. So $f(TOM, MTO)$ must be T. Consequently, by (3), $f(TMO, MTO) = T$.

Thus, so far we have shown that $f(TOM, MOT) = O$ and $f(TMO, MTO) = T$. We will now show that this leads to a contradiction.

We can repeat the argument that showed that $f(TOM, MTO) = T$ with M and T interchanged to show that $f(MOT, TMO) = M$. By interchanging the roles of Alice and Bob, we can similarly show that $f(TMO, MOT) = M$. By (3), this implies that $f(TMO, MTO) = M$, and this contradicts our finding that $f(TMO, MTO) = T$.

Example 1.4 (Path Selection) Suppose that there are three different paths R, S, T that can be used to transport traffic between two points (see Figure 1.1). The paths are to be used by two entities a and b, each of which has a preference relation over the paths (perhaps the paths have different performance characteristics which meet the needs of a and b differently). The network uses these

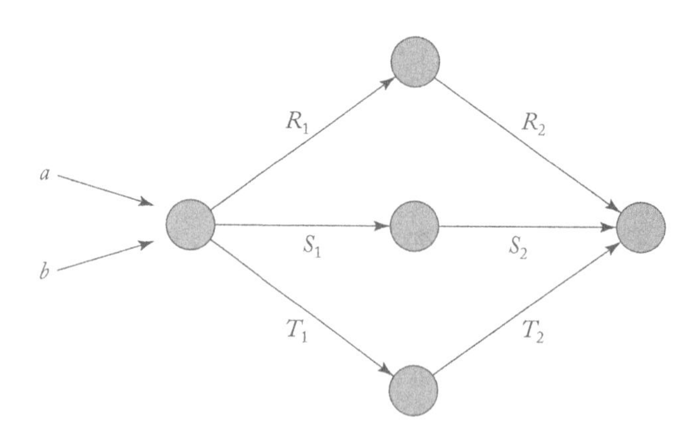

FIGURE 1.1: Path selection.

TABLE 1.1: The profile P

Voter 1	A	$>$	B	$>$	C
Voter 2	C	$>$	A	$>$	B
Voter 3	B	$>$	C	$>$	A

TABLE 1.2: The profile P'

Voter 1	A	$>$	B	$>$	C
Voter 2	C	$>$	A	$>$	B
Voter 3	C	$>$	A	$>$	B

preference relations to pick best path for a and b to share. This cannot be done while meeting conditions (1)–(3).

The consequences of Arrow's result are far-reaching. In addition to the impossibility of devising a reasonable SCF, it isn't possible to come up with a reasonable Social Welfare Function that outputs a globally fair ranking that combines the rankings of Alice and Bob. This further puts into doubt any plan for resource allocation that just depends on preferences.

VOTING

An election implements a social choice function. It maps the preferences of voters to a decision of who is elected. In view of Arrow's theorem, we may expect elections to have undesirable properties. We illustrate a few commonly used voting methods and we show that they are not monotone (i.e., they do not follow condition (3) of Arrow). For a fun and informative introduction to the issues around voting see [Poundstone 2008].

Example 1.5 (Simple Majority)
This intuitive scheme declares that a candidate is the winner if he is the first choice for more voters than any other candidate. But consider the profile P depicted in Table 1.1. Both A and B are contenders since they are first choice of one voter. Suppose we break the tie by declaring A the winner. If this scheme is monotone, then A should still be the winner in the profile P' shown in Table 1.2. Indeed, profile P' is obtained from profile P by improving the rank of A for the third

TABLE 1.3: A profile for the Borda count

No. Votes	
30	A > B > C
10	$C > A > B$
10	$B > C > A$
1	$A > C > B$
29	$B > A > C$
1	$C > A > B$

TABLE 1.4: Profile after candidate C drops out of the race

No. Votes	
$30 + 10 + 1 = 42$	$A > B$
$10 + 29 + 1 = 40$	$B > A$

voter. However, the simple majority rule declares C to be the winner for the profile P'. Thus, simple majority is not a monotone social choice function.

Example 1.6 (Borda Count)
For each vote, each candidate gets a score equal to the rank they received in that vote. A candidate's total score is obtained by adding their score over all the votes. The winner is the candidate with the lowest score. Although this seems like a reasonable strategy, it is not monotone. To see this consider the profile shown in Table 1.3. Note that candidate B wins this election. Assume that candidate C decides to drop out of the race. The resulting profile is shown in Table 1.4 and one can verify that candidate A is the new winner.

Example 1.7 (Instant Runoff)
In this iterative scheme, the voters first rank the candidates. Then, for each candidate, one counts the number of ballots for which he is ranked first. The candidate that has received the least number of voters rank as first is deemed the weakest and is eliminated. The procedure is repeated until only one candidate remains.

TABLE 1.5: Profile for instant runoff

No. Votes	
41	$A > B > C$
27	$B > A > C$
32	$C > B > A$

TABLE 1.6: Modified profile for instant runoff

No. Votes	
47	$A > B > C$
27	$B > A > C$
26	$C > B > A$
6	$A > C > B$

This scheme is non-monotonic, as one can see by considering the profile shown in Table 1.5. Clearly, B is eliminated in the first round and A wins. Now suppose that A manages to convert six votes from $C > B > A$ to $A > C > B$, so that the profile is now as shown in Table 1.6. Clearly, A should still win. But now observe that candidate C gets eliminated and A loses!

ANY LIGHT AT THE END OF THIS TUNNEL?

Notice that the SCF needs to map to an alternative for any possible choice of Alice and Bob's preferences. But some of these preferences may be absurd and therefore unreasonable for anyone to hold. For example, if applications express preference in terms of (rate, delay) pairs, then a real time video application is unlikely to ever prefer a higher delay for the same rate. It turns out that it is possible in some settings for Arrow's theorem to not hold when the set of possible preference relations is constrained in this way.

Another, more robust way out is to model preferences probabilistically. Given N alternatives, and strict preferences among them, there are $N!$ different preference relations possible, and rather than picking one for Alice, we might treat her preferences as a probability distribution over the $N!$

preference relations. This models the situation in which Alice's preferences can change, and also gets us out of the straight jacket of transitivity combined with conditions (1)–(3). It opens up a slew of possibilities on how to combine preferences. The fundamental ideas behind this formulation were developed by Diaconis and others, and this approach continues to be an active area of research. We will not explain this line of research here, but the interested reader is referred to [Diaconis 1989] for the fundamentals and [Jagabathula and Shah 2008] for an application.

1.2 UTILITIES

Given the problems with specifying and combining preferences, perhaps the situation would be simpler if each user were to assign a numerical value, called a *utility* to each option. As we discussed before, this only makes sense when the underlying preferences are transitive. Notice that specifying utilities gives us more than an ordering among options; it also indicates the degree to which one option is preferred over another.

COMBINING UTILITIES

Say that Alice assigns a utility $U_A(\cdot)$ to each activity. For instance,

$$U_A(movie) = 20, \qquad U_A(opera) = 12, \qquad U_A(tennis) = 5.$$

Assume that Bob's utilities are

$$U_B(movie) = 11, \qquad U_B(opera) = 15, \qquad U_B(tennis) = 24.$$

It may seem reasonable for them to choose the activity that maximizes the sum of their utilities, i.e., to go to a movie. However, a number of difficulties arise.

It may be that Alice and Bob use different scales to express their utility. The difference between Alice and Bob's first and second choices is 8 for both them, but perhaps Bob assigns utility on a linear scale while Alice operates on a logarithmic scale. This could imply that Bob is much less wedded to playing tennis than Alice is to going to a movie. Thus, it is far from obvious how to construct an individual utility function unless there are, at the very least, common units in which they can be measured.

Even when the units of utility are tied to money and willingness to pay, combining utilities may be problematic. Suppose Alice is much richer than Bob. She might be willing to pay much more for her top choice than Bob and maximizing the sum of the utility is effectively doing what Alice wants. Clearly, that isn't fair since Alice gets to be a dictator. If Alice were willing to pay Bob to get her way, (i.e., if utility can be transferred from one agent to another), he might agree, but (among other things) why should Bob be truthful about how much he needs to be paid to go along with Alice?

Thus, three considerations about utilities are as follows.

- Construction: Can each agent construct a utility function?
- Units: Can the utility functions from different applications be arithmetically combined, for example by adding? Is the utility like money?
- Transfer: Can utility from one agent be transferred to other agents?

If the answer to any one of these questions is negative, then it is difficult to find a firm basis to make rational choices.

In the context applications using a network, a possible way to define units might be in terms of performance. In particular, the network could mandate a specific set of application types, each of which has a predefined utility curve that is a function of throughput and delay, and that maps to a fixed range, say, $[0, 100]$. Each network session might then declare its type to the network, which allocates resources accordingly. But these units are not easily transferrable.

QUASILINEAR UTILITIES

To make progress, one needs to assume that utilities are *quasi-linear*. This means that they are *monetary* and *transferable*. Thus, the utility for Alice of activity x plus some quantity y of money is $U_{\text{Alice}}(x) + y$. In our previous example, this means that Alice is equally satisfied with playing tennis and getting \$15.00 as with watching a movie.

While quasilinear utilities are commonly used in the literature, applying them in the context of network resources has at least two difficulties.

- The monetary value of utility is questionable because money itself has a concave utility function. Indeed, an extra dollar in a person's pocket is worth more to him when he is poor than when he is rich. Thus, it is not fair to measure utility in dollars. The net effect of such a system would be that rich entities of the Internet would effectively look like dictators to the rest of user population. This brings up the important issues of Justice and Network Neutrality which are currently being debated.[2]

2. It could be argued that making inter-agent comparisons in the context of networking is simpler than in the welfare of people. However, it is important to realize that the Internet supports applications designed to be used by people. Thus, these applications will, in many instances, reflect the utility of individual application designers and users.

- Transferability is also questionable. Assume that the network tries to maximize the sum of the utilities of Alice and some other user Ted by letting Alice watch the movie in HD and shutting out Ted's connection. Can Alice really pay Ted for him to be satisfied with this solution? Alice may not even know Ted! Who collects the money and redistributes it?

MAXIMIZING USER WELFARE

Consider a set of possible allocations $\mathbf{x} = (x_i, i = 1, 2)$ of resources to two users $i = 1, 2$. Assume that user i receives a utility $U_i(x_i)$ from the allocation x_i. Consider the set F of possible pairs of utilities $(U_1(x_1), U_2(x_2))$, as shown in Figure 1.2. The functions $U_i(\cdot)$ are assumed to be strictly increasing and concave.[3]

The set of points on the "upper-right" boundary of that set are said to be *Pareto efficient*. These are allocations such that one cannot improve the utility of one user without decreasing that of the other. Thus, reasonable allocations must be Pareto efficient. However, choosing which of these allocations is desirable involves trading off the utility of one user for that of the other, and it is not obvious how to make that trade-off.

When utility is *transferrable* and agents can make side payments, the best allocation is to maximize the sum of the utilities, called the *user welfare*.[4]

In our example, Alice and Bob go to the movie and Alice gives \$4.50 to Bob. Another option is to go to the opera with Bob giving \$1.50 to Alice, and playing tennis with Bob giving \$9.50 to Alice. They both will prefer the first option since their utility is then \$15.50, larger than for the other possibilities.

In our example with two users,

$$\sum_i U_i(x_i) \leq \sum_i U_i(x_i^*), \tag{1.1}$$

where x^* is the allocation that results from maximizing the sum and x is another allocation. Now starting from x^* we can transfer utility to create new utilities u_i such that $u_i \geq U_i(x_i)$ for each i. Hence, when utility is transferable, the Pareto efficient pairs of allocations are no longer the upper-

3. A set is convex if it contains the line segment between any two of its points. A function f defined over a convex set is said to be concave if for any two members of the set x_1 and x_2:

$$f(\theta x_1 + (1-\theta)x_2) \geq \theta f(x_1) + (1-\theta) f(x_2)$$

for all $\theta \in (0, 1)$. The function is strictly concave if the inequality is strict. A function f is (strictly) convex if $-f$ is (strictly) concave.

4. This is true under mild assumptions on the utilities.

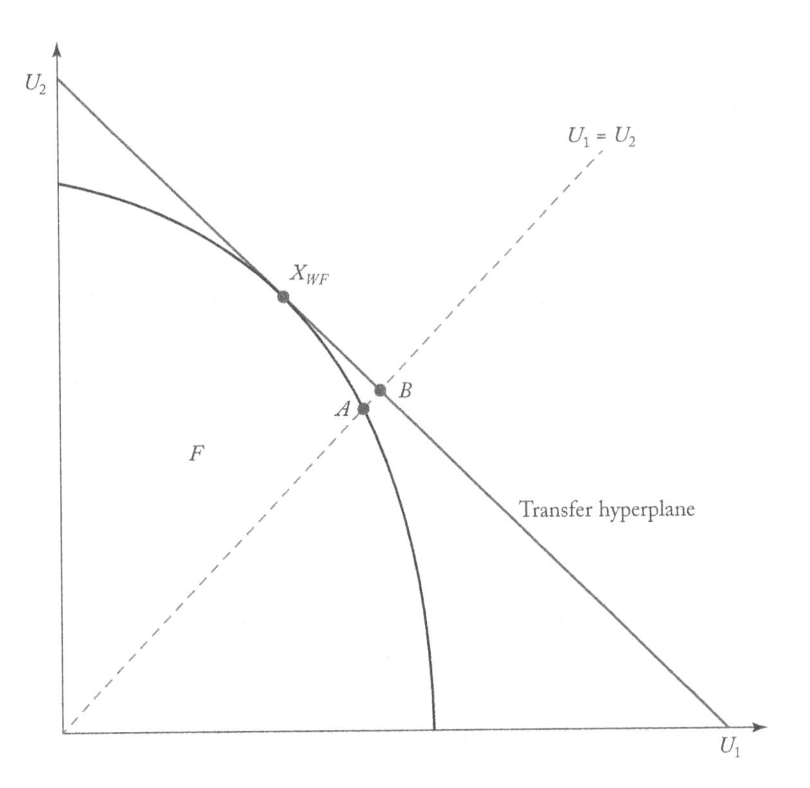

FIGURE 1.2: Suppose the goal is to find the best allocation for which $U_1(x_1) = U_2(x_2)$. If utility cannot be transferred, the best one can do is point A. The welfare maximizing point is given by X_{WF} and if utility is transferable, any point on the transfer hyperplane represents a different way in which this utility can be distributed. Notice that these distributions lie outside the original feasible region F, and in particular one an can achieve point B which is strictly better than A.

right boundary points of F, but are instead the points on the transfer hyperplane shown in the figure.

How does the Internet deal with application utility? Instead of Alice and Bob, consider Email and Movies, two applications that have very different utility functions. The surprising fact is that the Internet treats both applications as if they have the *same* utility function. Why does it do that and what are consequences? The simple reason the designers of the TCP protocol did this is for ease of implementation. Link bandwidth is plentiful enough that such a suboptimal approach works quite well. TCP acts to maximize the welfare of all the sessions on the network, assuming they all have the same utility function.

But this begs the question of what that utility function is. By modeling the behavior of TCP analytically it is possible to show heuristically that for a session that has a round trip delay of T, this function is approximately given by

$$U(x) = -\frac{1}{2T^2x},$$

where x is the rate received by the session. Given this function of utility what TCP implicitly does is to attempt to maximize the sum of the utilities of all the sessions on the Internet. See [Shakkottai and Srikant 2008] for the details of the analysis.

Notice that utility is only a function of rate and that origin-destination pairs that are close are favored over those that are far. Again, these limitations are introduced for implementation ease and abundant link capacity. However, when bandwidth becomes scarce, a more elaborate protocol may be required.

THE REPUGNANT CONCLUSION

When utility is *not transferrable*, maximizing the sum of all the utilities could result in strange allocations. For example, consider the thought provoking reasoning offered by Derek Parfit [Parfit 1986]. Let us represent the welfare of a society in two dimensions where the height represents utility level and the width represents population (see Figure 1.3).

Then under society A, everyone enjoys an excellent level of utility. Society A' has the same number of extremely happy people as A, but it has more people who are very happy. It seems clear that society A' is preferable to A because the total happiness has been increased, with no cost to the extremely happy. In society B, everyone is equally happy but the level of utility is lower than the

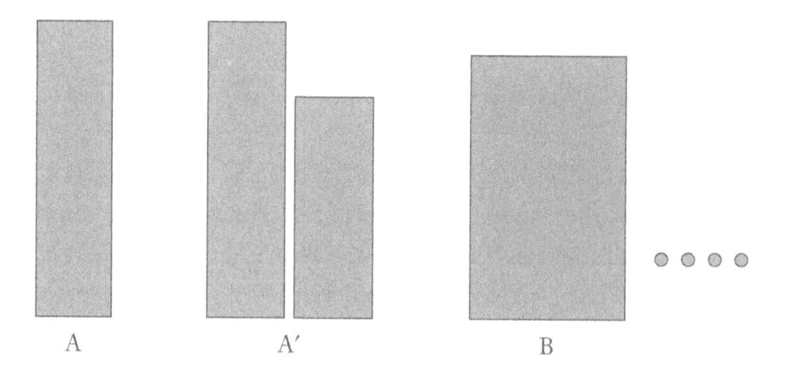

FIGURE 1.3: The repugnant conclusion.

happiest people in A'. Assuming that the average utility is higher in B than in A', we see that B is a more "equal" society with a net gain in average and therefore "total" happiness. Thus, B is a better society than A', implying that B is a better society than A. Now we can extend this argument by defining B' and C so that C will contain more people than B, its people are less happy, but we prefer it to B (and A). Continuing this process to some point (say Z), we have lots of people whose lives are just worth living and we prefer this society to the one in which everyone was extremely happy. Parfit called this the Repugnant Conclusion.

In the next section we will discuss a reasonable way of combining utilities due to Nash when they are not transferrable. We will require, however, that the agents make their decisions independently and there is no external agency that can compel to cooperate. Not surprisingly, this is called a *non-cooperative game*. It is not clear how to combine utilities in a *cooperative game*. (See [Shapley 1988] for Shapley's approach and [Peleg and Sudhölter 2007] for a more comprehensive treatment.)

THE NASH BARGAINING SOLUTION

Nash proposed a way of combining utilities that deals with the issues of Units and Transferability. He reasoned that two similar agents, r and s, might express their utilities differently, because they have different notions of units and the origin. That is, one may have $U_r(x_r) = aU_s(x_r) + b$, for some $a > 0$. Thus, Nash assumes a limited discrepancy in the nature of the units. The parameter a addresses linear utility inflation and b addresses the fact that different agents may have different incentives to even participate in the resource allocation process.

Nash sought to devise a way of combining utilities in a manner that treats agents r and s as if they have the same utility function. [5]

Nash [Nash 1950] suggests that rather than maximizing the sum of the utilities, we maximize their product (see Figure 1.4). More precisely:

$$NASH: \max_{x \geq 0} \Pi_{i=1}^{N}(U_i(x_i) - U_i(0)) \tag{1.2}$$

such that x is feasible.

NASH is insensitive to changes in units and origin. To see this, suppose each application r changes its utility function to:

$$U'_r(x_r) = a_r U_r(x_r) + b_r, \quad a_r > 0.$$

5. This clearly does not happen when maximizing welfare, i.e., $\sum_i U_i()$, since the optimal rate allocations will depend on a and b.

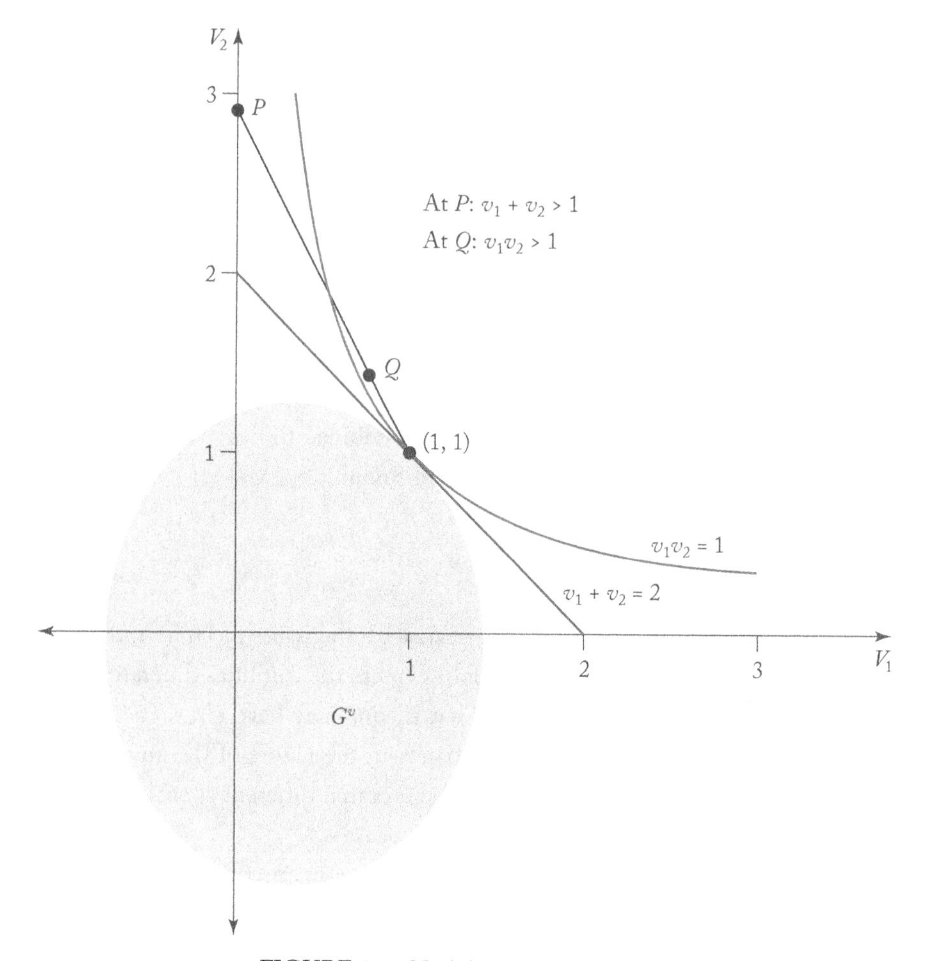

FIGURE 1.4: Nash bargaining solution.

We have: $U_r'(0) = a_r U_r(0) + b_r$. Thus,

$$\begin{aligned} U_r'(x_r) - U_r'(0) &= a_r U_r(x_i) + b_r - a_r U_r(0) - b_r \\ &= a_r (U_r(x_i) + U_r(0)). \end{aligned}$$

Letting $\alpha = a_1 a_2 \ldots a_n > 0$ the optimization problem (1.2) becomes:

$$\max_{x \geq 0} \Pi_{i=1}^n U'(x_r) - U_r'(0) = \alpha \Pi_{i=1}^n (U_i(x_i) - U(0))$$

$$\text{such that } x \text{ is feasible.}$$

Thus, the optimal rate allocations do not change with changes in units and origin.

The allocation resulting from *NASH* is called the Nash Bargaining Solution (NBS). It has a number of pleasing properties. Let x^* be the allocation that results from solving *NASH*. Then we have the following.

P1 Pareto Allocation. There is no feasible allocation $\hat{x} \neq x^*$ such that $\hat{x} \geq x^*$.

P2 Symmetry. The solution does not depend on how the applications are labeled. Any permutation π on the indices, results in a solution that applies π to the x^*.

P3 Independence of Irrelevant Alternatives. If x^* solves *NASH* over a feasible set, $\mathcal{F}$ and x^* is contained in a set $\mathcal{G} \subset \mathcal{F}$, then x^* should solve *NASH* over the feasible set $\mathcal{G}$ as well.[6]

Quite surprisingly, it turns out that *NASH* is the only way to combine utilities that is insensitive to units and origin, and that obeys $P1 - P3$.

Theorem 1.1 NBS is the only allocation that satisfies $P1$–$P3$ and independence from changes in units and origin.

Before proving the Theorem consider this example:

Example 1.8 (Dividing Money using NBS) Imagine that Alice and Bob want to split \$100.00. The question is how much each should get. The natural answer, \$50.00 each, may not be reasonable. To formulate the problem more precisely, say that Alice derives a utility $A(x)$ from getting x dollars and that Bob derives the utility $B(x)$. ($A(0) = B(0) = 0$). Then the NBS is to give x to Alice where x maximizes $A(x)B(100 - x)$. Assume that $A(x) = \sqrt{x}$ and $B(x) = x$, then one finds that $x^* = 33.3$. Under welfare maximization we give almost all the money, \$99.75 to Bob. Thus, when utility is not transferable, it seems much more reasonable to use the NBS rather than maximizing welfare.

Proof of Theorem 1.1. We have already verified that NBS is independent of changes in units and origin. It is easy to see that it follows P3 as well. We will show that it is the only solution to follow P1 and P2 as well.

For ease of exposition we assume only two applications—the extension to more than two is straightforward. Denote the convex and compact feasible region as E. Consider the plane defined by the axes $U_1(x_1)$ and $U_2(x_2)$. The feasible region defines a region in this plane $G = \{(U_1(x_1), U_2(x_2)) : (x_1, x_2) \in E\}$. This region is convex and compact since the U_i's are concave and E is convex. Now since the NBS is invariant to changes in units and origin, we can scale the axes so that the NBS is at the point (1, 1) and the transformed utility functions, $V_1(x)$ and $V_2(x)$ are such that the origin is contained in the transformed plane, G^v. If G^v is symmetric, then the only solution to meet P2 is of

6. This property states that "throwing out" non-optimal points from the feasible region does not change the solution.

the form (a, a). We now show that $(1, 1)$ is the only feasible point of the form (a, a) that is Pareto optimal to complete the proof.

First, we claim that all points $(v_1, v_2) \in G^v$ are such that $v_1 + v_2 \leq 2$. Note that since $(1, 1)$ maximizes $V_1(x)V_2(x)$:

$$v_1 v_2 \leq 1 \quad \text{for all } (v_1, v_2) \in G^V. \tag{1.3}$$

Now consider the line segment which connects the $(1, 1)$ to any point $(\bar{v}_1, \bar{v}_2)$ such that $\bar{v}_1 + \bar{v}_2 > 2$. This line segment consists of the points:

$$(w_1, w_2) = (\theta + (1-\theta)\bar{v}_1, \theta + (1-\theta)\bar{v}_2)$$

for $\theta \in [0, 1]$. We claim that there is a $\theta \in (0, 1)$ for which $w_1 w_2 > 1$, i.e.,

$$w_1 w_2 = \theta^2 + \theta(1-\theta)(\bar{v}_1 + \bar{v}_2) + (1-\theta)^2 \bar{v}_1 \bar{v}_2 > 1$$
$$\theta(1-\theta)(\bar{v}_1 + \bar{v}_2) + (1-\theta)^2 \bar{v}_1 \bar{v}_2 > 1 - \theta^2$$

For $\theta < 1$:

$$\theta(\bar{v}_1 + \bar{v}_2) + (1-\theta)\bar{v}_1 \bar{v}_2 > 1 + \theta$$
$$\theta > \frac{1 - \bar{v}_1 \bar{v}_2}{\bar{v}_1 + \bar{v}_2 - (1 + \bar{v}_1 \bar{v}_2)}.$$

Now we show that the RHS is in $[0, 1]$ if $(\bar{v}_1, \bar{v}_2) \in G^v$. We have $\bar{v}_1 \bar{v}_2 \leq 1$ and $\bar{v}_1 + \bar{v}_2 > 2$. Thus, both numerator and denominator are positive. Also,

$$1 - \bar{v}_1 \bar{v}_2 < \bar{v}_1 + \bar{v}_2 - 1 - \bar{v}_1 \bar{v}_2$$

implying that the RHS is less than 1. Thus, by picking a value of

$$\hat{\theta} \in \left(\frac{1 - \bar{v}_1 \bar{v}_2}{\bar{v}_1 + \bar{v}_2 - (1 + \bar{v}_1 \bar{v}_2)}, 1\right)$$

we can find a point (w_1, w_2) on the line segment connecting $(\bar{v}_1, \bar{v}_2)$ and $(1, 1)$ such that $w_1 w_2 > 1$. But since the feasible region is convex, and both end points of the line segment are feasible, (w_1, w_2) must be feasible as well. This contradicts (1.3) and proves our claim that $v_1 + v_2 \leq 2$ for all $(v_1, v_2) \in G^v$.

Now observe that the components of the NBS sum up to 2. Thus, it is a Pareto optimal point.

■

The NBS adjusts the U_r for scale and origin and then attempts to equalize the values of the resulting utility functions. Note the following relationship between NBS and Proportional Fairness.

Lemma 1.1 If $U_r(x_r) = a_r x_r + b_r$, $a > 0$, then the NBS is identical to the proportionally fair allocation.

1.3 SUMMARY AND REFERENCES

Social Choice Theory is a well-developed field within economics [Mas-Colell et al. 1995, Kreps 1988]. Its application to network resource allocation was pioneered by [Shenker 1995], and the idea of maximizing welfare was further explored in [Kelly 1997, Maulloo et al. 1998]. Utility is generally assumed to be transferrable, perhaps in the interest of making progress. For more on the issues around comparing utilities across individuals, see [Elster and Roemer 1991] for a glimpse into the range of views and [Harsanyi 1990, Binmore 2009] for specific views. The Nash Bargaining solution concept [Nash 1950] is sometimes called Proportional Fairness [Maulloo et al. 1998].

CHAPTER 2

Allocating Divisible Resources

Many network resources can be treated as divisible. For example, if two applications are sharing a link of capacity 1, then if user 1 gets to transmit at rate x_1 it is reasonable to allow x_1 to be any real number in the interval $[0, 1]$. The same is true for storage. In both cases, the unit at which things are divided (bit/sec or bit) is much smaller than the amounts that have to be allocated. In this chapter we explore various ways in which such resources can be shared among applications given their utilities and preferences.

2.1 IS SHARING A LINK LIKE SHARING A PIE?

Before setting about the task of allocating link resources it is important to point out that a link carrying packet switched traffic is a somewhat peculiar resource. Figure 2.1 depicts a typical scenario in which packets from several applications must be multiplexed on one link. Application packets typically arrive in bursts, so that during some intervals there may be more traffic than the link can immediately transmit, and this excess traffic is stored in a buffer. There may be other instances in which there is hardly any traffic and the buffer is mostly empty.

Let $\lambda = x_1 + x_2$ be the aggregate rate at which the application packets arrive and let r be the rate of the link. If user 1 is allocated x_1, it does not make sense to allocate more than $r - x_1$ to user 2 since that would make the buffer unstable. Thus, the state of the link at any given time must include the state of the buffer. For many systems in which packet arrivals are bursty, the average number of packets waiting in the buffer goes up exponentially as λ/r approaches 1. This behavior shows that without further simplifying assumptions, the link is too complicated to be modeled as a pie.

Treating a link as a queue, and the network as a system of interconnected queues can be useful but is very difficult to analyze [Kleinrock 1975, Walrand 1988, Parekh and Gallager 1994, Srikant and Ying 2014]. When link capacity is a scarce resource, as was the case in the early days of the Internet, network designers exploited the store-and-forward nature of packet switching to build large buffers to queue delayed packets.

Today's network designers build in smaller buffers (normalized by link speed) and use flow control to ensure that queues do not become too large and packet loss is rare. This is largely the role of TCP in the Internet. When queues do not become large, it makes much more sense to ignore

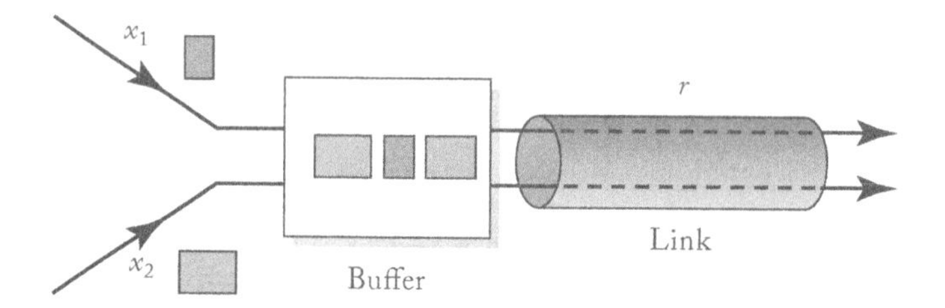

FIGURE 2.1: Two applications share a link with rate r. The link is equipped with a buffer to store packets before they are transmitted.

them and to treat a link like a pie—if user 1 gets x_1 then user 2 can get $r - x_1$. Of course restricting the ability for queues to form inside the network forces the buffering to occur outside of it, in the end devices.

Another aspect of packet transmission that must be considered is that it takes time for each packet to be transmitted, and clearly when one packet is being transmitted no packets from other applications can be transmitted at the same time. However, if the amount of time taken to transmit a packet is negligible (which is true for even reasonably fast links) then we can think of the traffic as a "fluid" which is infinitely divisible.

Mathematically, the fact that one may define feasibility conditions based on the rates of the flows and ignore the queueing phenomena is justified in Chapter 6. The key observation in that chapter is that, for reasonable models of queueing, the fluctuations around the mean rates do not accumulate and result in congestion collapse, as happens on highways. Instead, the queueing system behaves closely to what is predicted by considering only average rates.

2.2 DIVISIBLE LINKS

Consider a situation where two users share a link with rate 1. Say that user 1 gets to transmit at rate x_1 and user 2 at rate $x_2 = 1 - x_1$. How should we choose x_1? We examine a number of possibilities.

A "FAIR" SHARE

One sensible choice is $x_1 = 1/2 = x_2$. This may appear to be a fair allocation. Both users get to transmit at the same rate. However, is this really a good allocation?

MAXIMIZING USER WELFARE

Assume that the application of user 1 works perfectly as long as $x_1 \geq 0.8$ and does not work at all if $x_1 < 0.8$. Similarly, assume that user 2's application works perfectly as long as $x_2 \geq 0.1$ and does not work at all if $x_2 < 0.1$. Then, it is clear that one should choose $x_1 = 0.8$ and $x_2 = 0.1$. This allocation

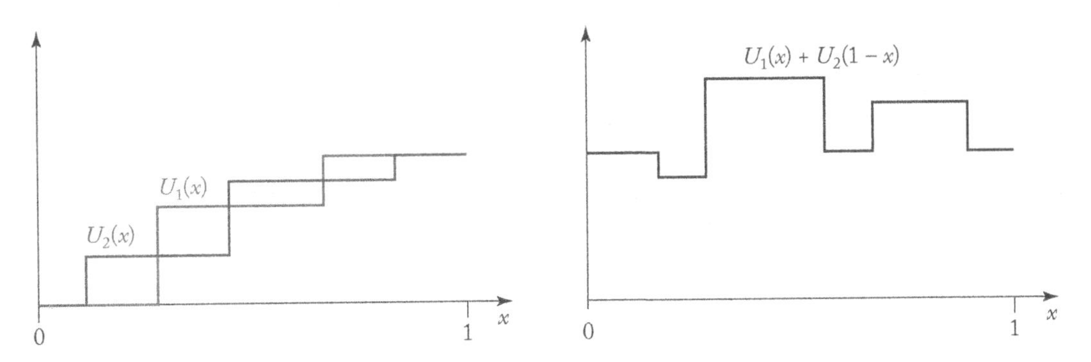

FIGURE 2.2: The utilities of the two users (left). The sum of utilities (right).

maximizes the utility of the two users, even though it does not seem fair. The sum of the user utilities is called the *user welfare*. An allocation that maximizes the user welfare is said to be *efficient*.

COMPLEXITY

Consider a situation where the utilities of the two users 1 and 2 are as shown as the functions $U_1(x)$ and $U_2(x)$, respectively, of their transmission rate x in the left-hand side of Figure 2.2. These utilities are non-decreasing, capturing the fact that many applications benefit from a larger transmission rate. The right-hand side of the figure shows the sum of the utilities when the two users share a link with rate 1 as a function of the rate x allocated to user 1. The point of the figure is to show that this sum can be a complicated function that does not have nice monotonicity or concavity properties. Finding the maximum of this function requires essentially to consider all possible values of x.

CONCAVE UTILITY FUNCTIONS

Figure 2.3 depicts the utility of a particular application as a function of *responsiveness*, defined as the reciprocal of the response time. When the network is very slow (responsiveness is close to zero), the utility is so small that it may make no sense to even use the application. As the responsiveness increases, utility rapidly increases until the responsiveness reaches some threshold, depicted in the figure. After that, there are *diminishing returns* of the increased responsiveness. Thus the utility function is *convex* from $[0, e]$ and *concave* for $x \geq e$. Diminishing returns means that adding one unit of responsiveness increases the utility less and less, which happens when the utility function is concave.

The property of diminishing returns of utility can be exploited in the design of resource allocation schemes. Intuitively, if one is maximizing the sum of the utilities (the user welfare), it makes sense to take away resources from applications that are operating at flatter portions of their

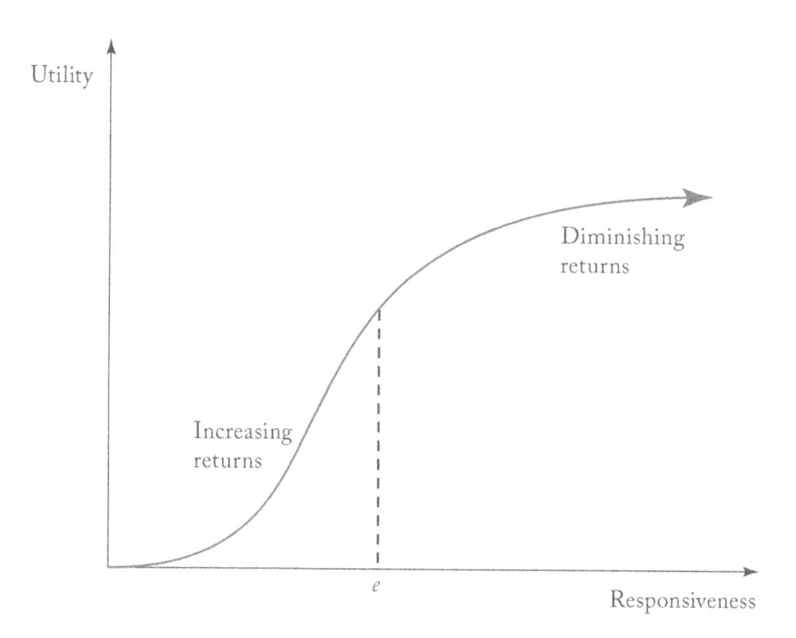

FIGURE 2.3: A typical utility function.

utility curve and to give them to applications that operate on a steeper portion. Such reallocations increase the total utility without severely penalizing a few applications.

There is, however, a set $[0, e]$ of values of the responsiveness for which the utility is convex and does not exhibit diminishing returns. If the network is fast enough, there is a negligible probability of being in this regime, and it is reasonable to treat utility as concave. The implicit assumption here is that if the network is not fast enough to meet the threshold e where diminishing returns start occurring, then that application is not likely to be deployed at significant levels. Similar considerations apply when the performance measure is the transmission rate instead of the responsiveness.[1] These applications are ignored until the network "becomes faster." Thus, it is customary to assume that when application requirements are described as utility functions, these functions are strictly concave.

The shape of an agent's utility curve can also be interpreted as their attitude towards risk. A concave utility denotes risk aversion. To see this suppose an application has the choice of getting a

1. Internet video has a threshold for the transmission rate that is approximately equal to 400 kb/s. In the early 1990s there were various video applications deployed but they did not get wide spread traction since performance rarely met the 400 kb/s threshold. As speeds have increased and broadband deployments have grown, the threshold is almost always met, and so the utility function of video has effectively become strictly concave.

rate of 50 Kbps with probability 1 and 100 Kbps with probability 0.5 (and 0 Kbps with probability 0.5). Then the utility of the first option is $U(50)$ and the expected utility of the second option is $0.5U(100)$. If $U(x)$ is concave, the expected utility is always higher if the application takes the first option. The degree of curvature in the function measures the degree of risk averseness. This is sometimes denoted by the Arrow-Pratt measure of absolute risk aversion:

$$r_U(x) = -\frac{U''(x)}{U'(x)}. \tag{2.1}$$

When $U(x)$ is linear, $r_U(x) = 0$ and the function represents risk neutral behavior.

GRADIENT PROJECTION

If one makes the assumption that utilities are concave, the situation is greatly simplified since the sum of concave functions is concave. For example, for two users, sharing a single link of capacity 1, one has to maximize the concave function $U_1(x) + U_2(1 - x)$. This can be done by a "greedy" algorithm. Such an algorithm cannot get trapped in a local maximum.

Generalizing, suppose that one has a network of users with various concave user utility functions as well as constraints that come from the link capacities. Then these constraints form a convex feasible region. We would like to maximize the sum of the utilities subject to the constraints.

Say that we want to maximize a concave differentiable function $g(x)$ (which in this case is the sum of the utilities) defined on a closed convex set $\mathcal{C}$ in $\Re^d$ (this is given by the constraints). Then as long as $g(x)$ does not change too rapidly, a simple algorithm called the *Gradient Projection Algorithm* can be used to find the maximizing allocation. In particular, suppose the gradient $\nabla g(x)$ of the function is *Lipschitz*[2], i.e., that

$$||\nabla g(x) - \nabla g(y)|| \leq K||x - y||, \forall x, y \in \mathcal{C}.$$

Then, the following algorithm converges to the set of maxima of $g(x)$ on $\mathcal{C}$:

$$x(n+1) = \{x(n) + \frac{1}{n}\nabla g(x(n))\}_{\mathcal{C}}, \quad n \geq 1.$$

In this expression, $\{z\}_{\mathcal{C}}$ denotes the *projection* of $z \in \Re^d$ on $\mathcal{C}$, i.e., the closest point in $\mathcal{C}$ to z. See [Boyd and Vandenberghe 2004] for a presentation of this result.

Clearly, the rate at which the updates converge to the maximum depends on the nature of $g(x)$.

2. The Lipschitz condition limits the rate at which a function can change. In particular, the absolute value of the slope of the line connecting any two points is bounded by a constant K.

MAX-MIN, MAX-SUM, NASH BARGAINING

Back to the two-user example, suppose that the utility of user 1 is $U_1(x_1) = \log(1 + x_1)$ and that of user 2 is $U_2(x_2) = \sqrt{x_2}$. These are two concave increasing functions. To maximize the minimum of the two utilities $\min\{U_1(x), U_2(1 - x)\}$, one chooses $x \approx 0.71$ (found numerically); this is the *max-min* allocation and it corresponds to $U_1 = U_2 \approx 0.54$.

To maximize the sum $\log(1 + x) + \sqrt{1 - x}$ we should choose $x = 0.464$; this is the *max-sum* allocation and it corresponds to $U_1 = 0.38$ and $U_2 = 0.732$.

As we explained in Section 1.2, Nash argued that one should maximize the *product* $\log((1 + x) \times \sqrt{1 - x})$, which occurs with $x = 0.61$. One calls this allocation the *Nash bargaining equilibrium* or *proportionally fair allocation* and it corresponds to $U_1 = 0.48$ and $U_2 = 0.62$.

The three different allocations are shown in Figure 2.4.

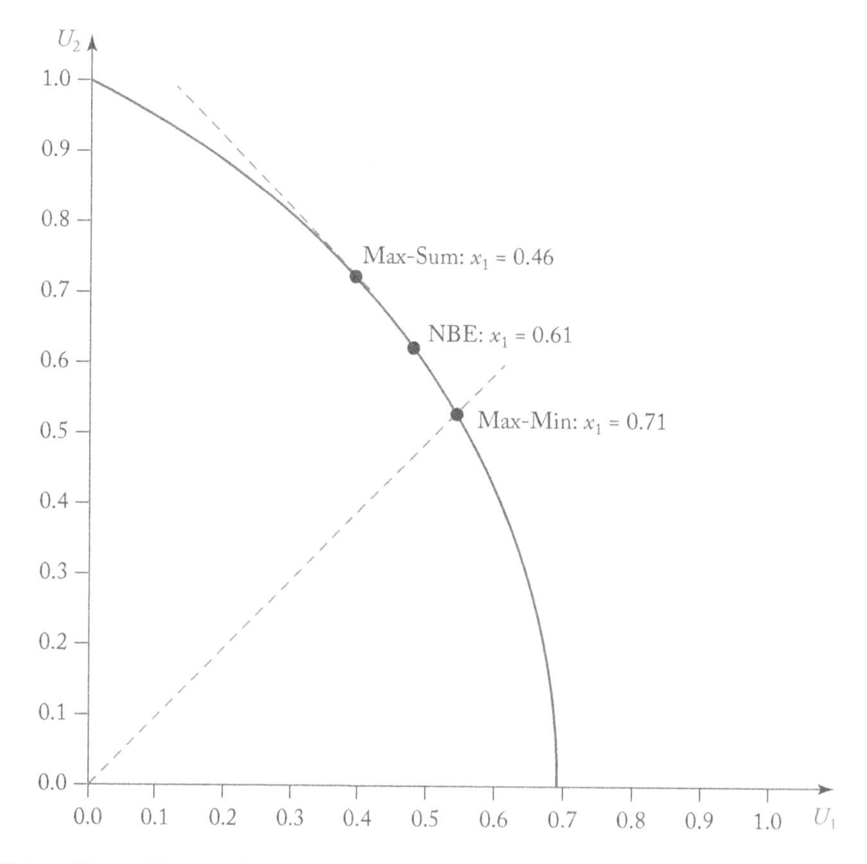

FIGURE 2.4: Three different allocations: max-min, max-sum, and Nash bargaining equilibrium (NBE).

The point of this example is that the allocation depends rather strongly on the objective, which is not surprising. Unfortunately, it is not easy to guess the utilities of users as they depend on the applications. In the Internet, the allocation does not depend on the application.

α-FAIRNESS

Even if one accepts the axioms that justify the Nash bargaining equilibrium for sharing resources among a set of users, one can argue that other objectives may be more appropriate. For instance, consider once again the case of two applications that share one link with rate 1 and with utilities $U_1(x)$ and $U_2(1-x)$. Imagine that, half of the time, Alice uses application 1 and is allocated rate x and Bob uses application 2 and is allocated rate $1-x$. Assume also that the rest of the time, the situation is reversed. The average utility that Alice and Bob face is then $(U_1(x)+U_2(1-x))/2$. Thus, both Alice and Bob are better off if one chooses x to maximize the sum of the utilities $U_1(x)+U_2(1-x)$, instead of maximizing the product $U_1(x)U_2(1-x)$. Another viewpoint is that Alice and Bob occasionally use the applications 1 and 2 and become very unhappy if one application does not perform well. In this case, they may leave the network provider if the minimum of $U_1(x)$ and $U_2(1-x)$ is not satisfactory. These considerations show that one may have to trade-off between the sum, the product, and the minimum of the utilities of different applications.

The three different problems max-min, max-sum, and Nash bargaining equilibrium can be unified into a single one. Define, for $u, \alpha > 0$,

$$\phi(\alpha, u) = \begin{cases} \frac{1}{1-\alpha} u^{1-\alpha} & \text{if } \alpha \neq 1 \\ \log u & \text{if } \alpha = 1. \end{cases}$$

Then the maximization of

$$\phi(\alpha, U_1(x_1)) + \phi(\alpha, U_2(x_2)),$$

called the α-*fair allocation*, is the maximization of the sum of the utilities when $\alpha = 0$, the maximization of their product when $\alpha = 1$, and one can show that it is essentially the maximization of their minimum when α is very large. Thus, one needs to consider only the maximization of the sum $\sum_i \phi(U_i(x_i), \alpha)$ of the modified utilities. The choice of α captures the tradeoff between *efficiency*, i.e., maximizing user welfare, and protecting the worst-off user (max-min). See [Mo and Walrand 2000].

STRATEGYPROOF?

Not surprisingly, if one tries to maximize the sum or the product of the utilities, the users have an incentive to lie about their utilities. For instance, assume that the utilities of the users are indeed

$\log(1+x_1)$ and $\sqrt{x_2}$ but that user 1 lies about his utility and declares that it is $\log(x_1)$ instead of $\log(1+x_1)$. The max-sum allocation is now $x_1 = 0.83$ instead of 0.48, the Nash bargaining equilibrium is now $x_1 = 1$ instead of 0.61, and the max-min allocation is $x_1 = 1$ instead of 0.71. An allocation mechanism is said to be *strategyproof* or *incentive-compatible* if users have no incentive to lie. We discuss this issue in Section 2.6.

PRICING AND REVENUE

From our daily experiences, it is clear that setting prices for resources can be a powerful way to allocate resources. Say that the link charges a price λ per unit bandwidth. User i then chooses x_i to maximize $U_i(x_i) - \lambda x_i$. In our example, user 1 chooses x_1 to maximize $\log(1+x_1) - \lambda x_1$, so that $x_1 = \max\{0, 1/\lambda - 1\}$ and user 2 chooses x_2 to maximize $\sqrt{x_2} - \lambda x_2$, so that $x_2 = 1/(4\lambda^2)$. For a link price of $\lambda = 0.683$, one finds that $x_1 = 0.464$ and $x_2 = 1 - x_1$ thus meeting the capacity constraint of $x_1 + x_2 \leq 1$. Also, the resulting allocation agrees with the max-sum allocation.

There are two ways to view pricing. First, the price λ can be used as a coordination signal that leads to an allocation that maximizes the user welfare. We can modify the utilities as in our discussion of α-fairness so that this pricing scheme can be used to solve for the max-min fairness or the Nash bargaining equilibrium.

Second, one can view pricing as a mechanism to maximize the provider's revenue. In this case, the provider actually charges λ per unit of rate to each user. To simplify the discussion, let us consider the case of a single user with utility $U(x)$. If the provider charges a unit price λ, the user chooses the value $x(\lambda)$ of x that maximizes $U(x) - \lambda x$. The provider then chooses the value λ_P of λ that maximizes his revenue $R(\lambda) = \lambda x(\lambda)$ subject to $x(\lambda) \leq C$ if the rate of the link is C. Also, let λ_U be the value of λ that maximizes $U(x(\lambda))$ subject to $x(\lambda) \leq C$. Thus, λ_P is the price that the provider wants to charge and λ_U is the price that results in the maximum user welfare.

As a first example, assume that $U(x) = a\log(1+x)$, so that $x(\lambda) = \max\{0, a\lambda^{-1} - 1\}$ and $R(\lambda) = \max\{0, a - \lambda\}$. In this case, $R(\lambda)$ is decreasing in λ and so is $x(\lambda)$. Hence, $\lambda_P = \lambda_U$. In this particular case, the interests of the user and of the provider are aligned.

However, these interests aren't always aligned. Consider a second example for which $U(x) = ax(1+x)^{-1}$. This case is illustrated in Figure 2.5 for $a = 6$. Then we find that $x(\lambda) = \max\{0, \sqrt{a/\lambda} - 1\}$ and $R(\lambda) = \max\{0, \sqrt{a\lambda} - \lambda\}$. The maximum value of $R(\lambda)$ occurs for $\lambda = a/4$. Also, $x(\lambda) \leq C$ if $\lambda \geq a(1+C)^{-2}$. Hence, we find that

$$\lambda_P = \begin{cases} a(1+C)^{-2} & \text{if } C \leq 1 \\ a/4 > a(1+C)^{-2} & \text{if } C > 1. \end{cases}$$

Thus, the interests of the provider and user are misaligned if $C > 1$. In this case, the provider charges a price that results in some unused capacity, creating some artificial scarcity.

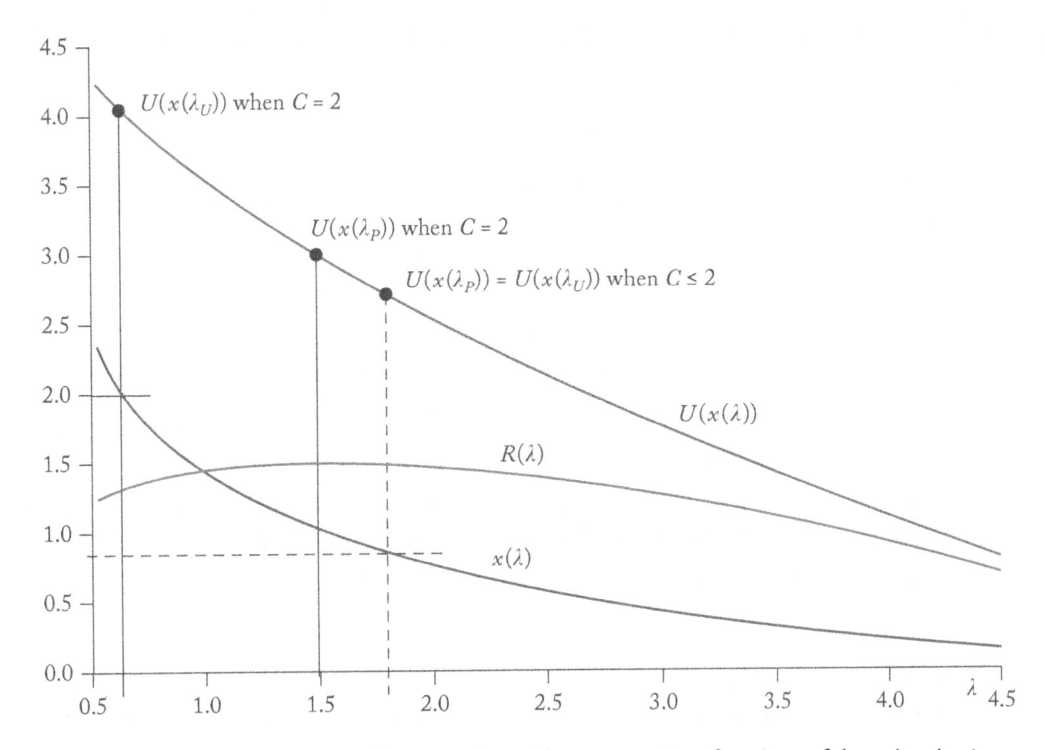

FIGURE 2.5: Rate x, user utility U, and provider revenue R as functions of the unit price λ.

Competition among providers would reduce this misalignment, and it is interesting to note that even a little bit goes a long way [Gajic et al. 2009].

TÂTONEMENT

Computing the price λ to maximize the user welfare requires knowing the two utility functions. But these utility functions are often not known. One way to get around this problem is to implement a scheme known as a *tâtonement process* (French for "groping in the dark"). In this scheme, the price λ gets adjusted down if $x_1 + x_2 < 1$ and up if $x_1 + x_2 > 1$, so that demand eventually equals supplies.

Note that this process is distributed: the price adjustment is done without knowing the utilities of the users and the users select their quantity only by looking at the price, not worrying about other users.

BIDDING

In yet another scheme, the two users bid some amounts b_1 and b_2 and the link capacity (equal to 1) is allocated in proportion to the bids. Thus, user i gets $x_i = b_i/(b_1 + b_2)$ and pays b_i (for $i = 1, 2$).

Assuming that $U_1(x) = 1 + \log(x)$ and $U_2(x) = 2\log(x)$, we find that user 1 gets a net utility (i.e., utility minus price)

$$1 + \log\left(\frac{b_1}{b_1 + b_2}\right) - b_1$$

and user 2 gets a net utility

$$2\log\left(\frac{b_2}{b_1 + b_2}\right) - b_2.$$

For a given bid b_2, the bid b_1 that maximizes user 1's net utility is some $b_1(b_2)$. Similarly, for a given b_1, the bid b_2 that maximizes user 2's net utility is some $b_2(b_1)$. One can check that $b_1 = 2 - \sqrt{2}$ and $b_2 = 3 + \sqrt{2}$ meets those two conditions. These bids correspond to the allocations $x_1 \approx 0.117$ and $x_2 \approx 0.883$ and to a sum of utilities $U_1(x_1) + U_2(x_2) \approx -1.39$, instead of the maximum possible value -0.91. We discuss further the effect of strategic users in Section 2.6.

VCG AUCTION

Yet another scheme uses an auction. In this scheme, the users bid, then the auctioneer calculates an allocation (x_1, x_2) and prices (p_1, p_2) based on the bids. The goal of the auction is to maximize the sum of the utilities $U_1(x_1) + U_2(x_2)$. The users, knowing the rules of the auction, choose the bids that maximize their net utility, i.e., the utility minus the price. We have seen that the tâtonement scheme achieves this goal, but in some situations a back-and-forth mechanism is not practical and the allocation has to be decided in one shot and not modified later.

Here is one strategyproof auction mechanism, called VCG (for Vickrey, Clark, and Grove). The two bidders announce utility functions $V_1(\cdot)$ and $V_2(\cdot)$ (they may lie and not reveal their actual utility functions). The auctioneer computes the allocation (x_1^*, x_2^*) that maximizes $V_1(x_1) + V_2(x_2)$ subject to $x_1 + x_2 \leq 1$. He then computes the prices as follows. The price p_1 is the reduction of utility of user 2 caused by user 1. That is, $p_1 = V_2(1) - V_2(x_2^*)$. Similarly, $p_2 = V_1(1) - V_1(x_1^*)$. Note that the net utility of user 1 is $U_1(x_1^*) - p_1 = U_1(x_1^*) + V_2(x_2^*) - V_2(1)$ where (x_1^*, x_2^*) maximizes $V_1(x_1^*) + V_2(x_2^*)$. Thus, user 1 is better off by revealing his actual utility function $U_1(\cdot)$ instead of some $V_1(\cdot)$, no matter what user 2 declares. Indeed, $U_1(x_1^*) + V_2(x_2^*)$ is maximized when the auctioneer knows the function $U_1(\cdot)$, because the auctioneer then chooses (x_1^*, x_2^*) to maximize $U_1(x_1) + V_2(x_2)$ instead of $V_1(x_1) + V_2(x_2)$. The same consideration applies to user 2. Hence, this mechanism is strategyproof and the users reveal their true utility functions that the auctioneer maximizes, thus resulting in the maximization of the user welfare.

We explain the VCG mechanism in more generality in Section 3.2.

2.3 MULTIPLE RESOURCES

There are many examples where users require more than one resource. For example, processing a job in a server requires some CPU cycles and some network capacity. Say that to process a job at rate x_i for user i, the system uses a rate $A(1, i)x_i$ of its CPU and a rate $A(2, i)x_i$ of its disk interface. There are two users and the feasible processing rates are such that $A\mathbf{x} \leq \mathbf{c}$ where $\mathbf{c} = (c_1, c_2)'$ and c_1 is the CPU rate and c_2 the capacity of the disk interface. Assume as before that user i gets a utility $U_i(x_i)$.

DUAL ALGORITHM

The example of pricing a single link can be generalized to multiple resources. Consider the following constrained optimization problem:

$$\text{Maximize} \sum_{i=1}^{n} U_i(x_i)$$

$$\text{s.t. } A\mathbf{x} \leq \mathbf{c} \text{ and } \mathbf{x} \geq 0, \tag{2.2}$$

where the U_i are concave differentiable functions, A is a non-negative $m \times n$ matrix and $c \in \Re^m$. Thus, there are m resources and each unit of x_i uses $A(j, i)$ units of resource j. There is a total quantity c_j of resource j.

We introduce a Lagrangian:

$$L(\mathbf{x}, \lambda) = \sum_{i=1}^{n} U_i(x_i) - \lambda'(A\mathbf{x} - \mathbf{c}),$$

where $\lambda \in \Re^m_+$ and λ' denotes the transpose of λ. The interpretation is that the constraint on each resource j in the optimization problem is replaced by a unit resource price λ_j. To solve the optimization problem, one maximizes the Lagrangian over $\mathbf{x}$ and then minimizes it over λ. To maximize over $\mathbf{x}$, each user i chooses x_i to maximize

$$U_i(x_i) - (\lambda' A)_i x_i. \tag{2.3}$$

Note that $(\lambda' A)_i = \sum_j \lambda_j A(j, i)$ is the price per unit of x_i since that unit consumes $A(j, i)$ units of resource j and that resource has a unit price λ_j. As in the single link example, we expect that by adjusting the prices, one will meet the constraints exactly.

To adjust the prices, one decreases the price of an under-used resource and one increases that of an over-used resource. For instance, one can use the following gradient projection algorithm:

$$\lambda(n+1) = \{\lambda(n) + \frac{1}{n}[A\mathbf{x}(n) - \mathbf{c}]\}_+.$$

LYAPUNOV APPROACH

Instead of the dual algorithm, an alternative analysis is based on the drift of a *Lyapunov function* for the system. This approach is easier to implement than the dual algorithm because the prices are proportional to the values of some counters. In a general discrete-time stochastic model, the method does not converge to the optimal utility, but the error can be bounded and made as small as desired. These two methods provide complementary ways to design dynamic resource allocation schemes.

To explain this method, we introduce counters that accumulate the requests for resources. As the users place requests at rate $\mathbf{x}$, the counters of the resources increment at rate $A\mathbf{x}$ and they decrement at rate $\mathbf{c}$. Let us denote the values of the counters by $\mathbf{q}$. To keep the system stable, we want to make sure that the counter values $\mathbf{q}$ remain small. This requires limiting $\mathbf{x}$. However, we want to achieve large utilities, which is a conflicting goal. To choose a good tradeoff between these goals, we choose $\mathbf{x}$ to maximize

$$\sum_i U_i(x_i) - \alpha \frac{d}{dt} V(\mathbf{q}(t)),$$

where

$$V(\mathbf{q}) = \frac{1}{2} \sum_j q_j^2$$

and α is a positive constant. The intuition for this choice is that one wishes the first term to be large and also the second one to be large to make the counter values decrease. Now,

$$\begin{aligned}
\frac{d}{dt} V(\mathbf{q}(t)) &= \frac{d}{dt} \frac{1}{2} \sum_j q_j^2(t) \\
&= \sum_j q_j(t) \frac{d}{dt} q_j(t) = \sum_j q_j(t)[A\mathbf{x} - \mathbf{c}]_j.
\end{aligned}$$

Thus, one chooses the value of $\mathbf{x}$ that maximizes

$$\sum_i U_i(x_i) - \alpha \sum_j q_j(t)[A\mathbf{x}]_j = \sum_i U_i(x_i) - \alpha \mathbf{q}' A\mathbf{x}. \tag{2.4}$$

Comparing with (2.3), we see that this approach corresponds to using

$$\lambda_j = \alpha q_j(t).$$

To estimate the performance of this control scheme, let x^* be the solution of (2.2). We have

$$\begin{aligned}\sum_i U_i(x_i(t)) - \alpha\frac{d}{dt}V(\mathbf{q}(t)) &= \sum_i U_i(x_i(t)) - \alpha\sum_j q_j(t)[A\mathbf{x}(t) - \mathbf{c}]_j \\ &\geq \sum_i U_i(x_i^*) - \alpha\sum_j q_j(t)[A\mathbf{x}^* - \mathbf{c}]_j \\ &\geq \sum_i U_i(x_i^*),\end{aligned}$$

where the first inequality comes from the fact that $x(t)$ maximizes (2.4) and the second from the fact that $A\mathbf{x}^* \leq \mathbf{c}$.

Integrating this inequality over $[0, T]$ and dividing by T, we conclude that

$$\frac{1}{T}\int_0^T \sum_i U_i(x_i(t))dt - \alpha\frac{1}{T}[V(\mathbf{q}(T)) - V(\mathbf{q}(0))] \geq \sum_i U_i(x_i^*).$$

Letting $T \to \infty$, we find

$$\liminf_{T\to\infty} \frac{1}{T}\int_0^T \sum_i U_i(x_i(t))dt \geq \sum_i U_i(x_i^*)$$

so that this control achieves the maximum performance.

In practice, the control is not in continuous time. A similar analysis in discrete time shows that the control achieves a performance at least equal to $\sum_i U_i(x_i^*) - \alpha B$, where B is a constant that depends on the maximum link rates and the time step of the control.

Example 2.1 (Two Users) As an illustration, assume as before that $U_1(x) = \log(1 + x)$ and $U_2(x) = \sqrt{x}$. Say that $A(1, 1) = 1$, $A(2, 1) = 2$, $A(1, 2) = 1$, $A(2, 2) = 3$ and $c_1 = 1$, $c_2 = 2$. Say that the demands at step n of the tâtonement scheme are $x(n) = (x_1(n), x_2(n))'$ when the prices are $\lambda_1(n)$ and $\lambda_2(n)$. We then update the prices as follows:

$$\lambda_j(n + 1) = \max\{0, \lambda_j(n) + \frac{1}{n}[A(j, 1)x_1(n) + A(j, 2)x_2(n) - c_j]\}.$$

That is, if resource j is over-subscribed at step n, one increases its price; otherwise, one lowers it. The step size in the algorithm is $1/n$ and decreases for the algorithm to converge to the optimum. At each step, the users choose the demands that maximize their net utility. As before,

$$x_1(n) = \max\left\{\frac{1}{p_1(n)} - 1, 0\right\} \text{ and } x_2(n) = \frac{1}{4p_2^2(n)}$$

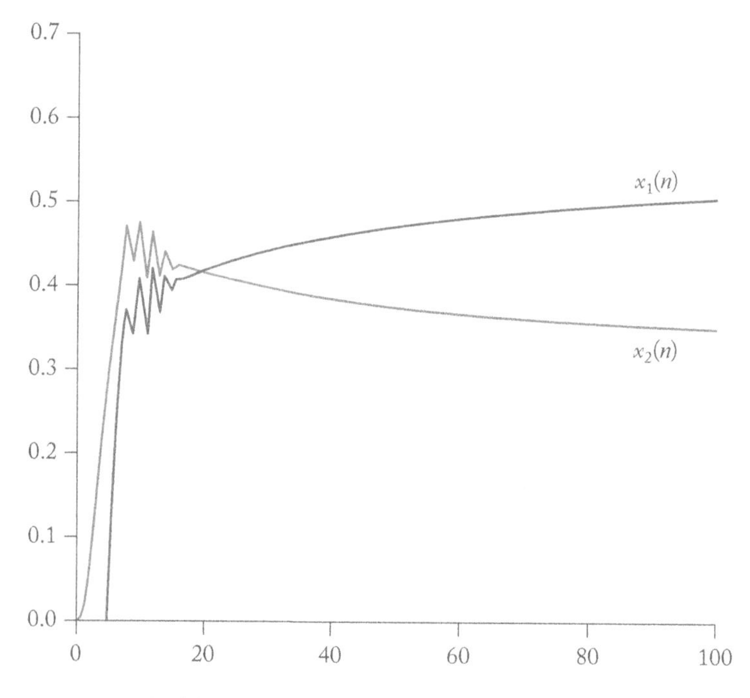

FIGURE 2.6: Convergence of dual algorithm.

where

$$p_i(n) = \lambda_1(n)A(1, i) - \lambda_2(n)A(2, i).$$

We call this procedure the *dual algorithm*.

Figure 2.6 shows that this scheme converges to the optimal allocation of the two resources.

2.4 SPECTRUM AS A RESOURCE

In Section 2.1, we discussed how different connections can share a link. Implicit in this system is the fact that sources are attached to a switch or router using different physical links. These sources send flows of packets that share one outgoing link of the device. In this section, we explore the situation in which a set of sources communicate with a set of receivers wirelessly. Each source has a transmitter that can be viewed conceptually as coding messages and modulating them in some fashion, and at some power. Since the power of a signal weakens with distance, a given receiver may receive signals with different strengths, and it may become difficult to decode the signal of interest. This phenomenon of interference makes the problem of sharing a wireless channel difficult. Let us consider three typical cases of asymmetry as depicted in Figure 2.7.

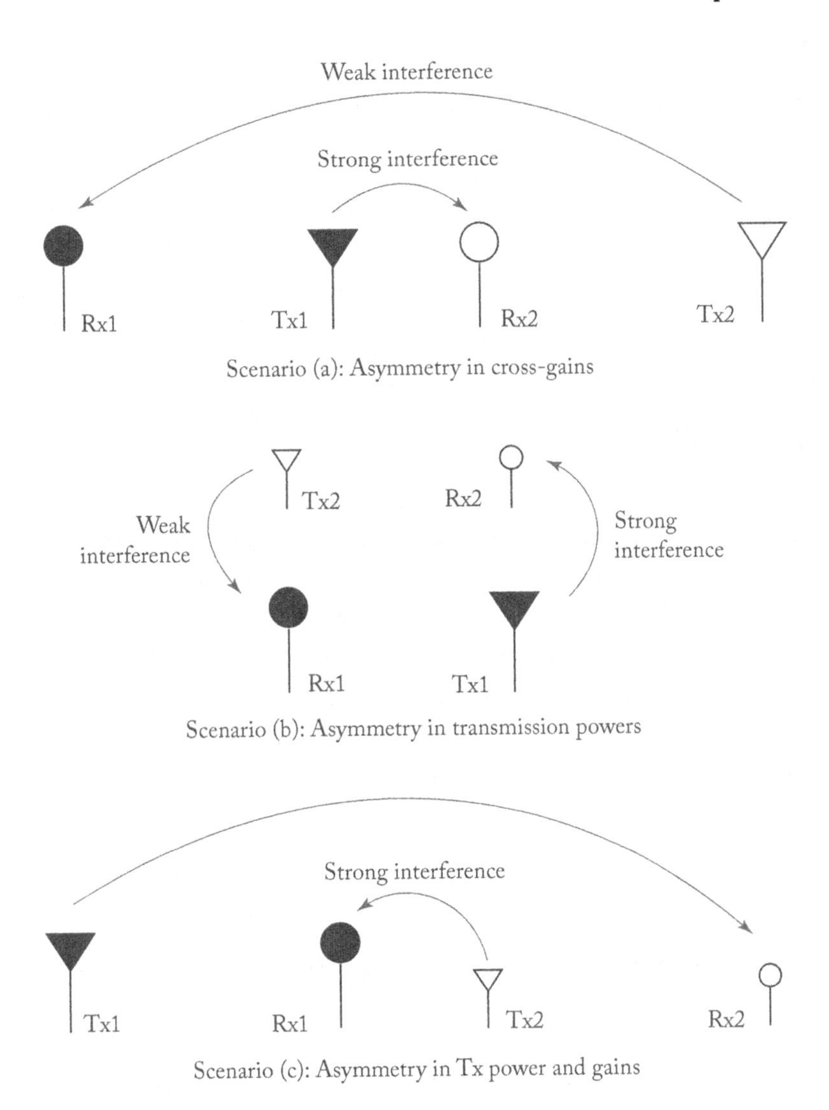

FIGURE 2.7: Three examples of asymmetric situations between two systems sharing the same band. The sizes of the antennas represent power capabilities, and smaller distances indicate higher gains.

There are two systems, each with one transmitter and one receiver. The heights represent the magnitude of the power. Thus in the top-most scenario both systems transmit with the same power, while in the bottom-most figure system 1 transmits with more power than system 2. Also depicted in the scenarios is the effect of placement. For example, in the top-most scenario, system 1 interferes with system 2 more than system 2 does with system 1 even though they both have the same transmit powers.

What does it mean to "allocate" the resource in this setting? One possibility is the following.

- Suppose that frequencies are limited to a range W Hz-wide.
- Allocate to each source a function $P_i(f)$ which is the maximum power at which the source i is allowed to transmit at frequency f.
- One special case of this is Frequency Division Multiplexing, in which the sources transmit over non-overlapping bands thereby eliminating all interference, and another is Full Spread in which source distributes their power equally over all bands.

Let's look at what happens if we have two systems and make some more simplifying assumptions:

TWO-SYSTEM EXAMPLE

We model a situation in which two systems, each formed by a single transmitter-receiver pair, coexist in the same area. Assume that the channel is a 2-user Gaussian interference channel in discrete time, [3] and that user i's input power is limited through an average power constraint, P_i. Two more simplifying assumptions: all interference is treated as noise (interestingly, sophisticated receivers can sometimes use strong interference to their advantage), and both users use Gaussian codebooks, so that the transmitted signals look like white Gaussian processes.

Under these assumptions, using the capacity expression for the single user Gaussian channel, we can determine the maximum rate that system i can achieve for specific power allocations (see [Tse and Viswanath 2005]):

$$R_i = \int_0^W \log\left(1 + \frac{c_{i,i} p_i(f)}{N_0 + \sum_{j \neq i} c_{j,i} p_j(f)}\right) df, \tag{2.6}$$

where $p_i(f)$ is the power spectral density of the input signal of system i, and where for convenience we defined $c_{i,j} = |h_{i,j}|^2$. Note that due to the power constraints, $p_i(f)$ must satisfy:

$$\int_0^W p_i(f) df \leq P_i. \tag{2.7}$$

The spectrum sharing problem that we consider is to determine a set of power allocations $\{p_i(f)\}$ for the two systems, that maximizes a given global utility function while satisfying the power constraints. As we have discussed earlier, a suitable choice of utility functions will allow us to

---

3. This channel is defined by:

$$y_i[n] = \sum_{j=1}^{2} h_{j,i} x_j[n] + z_i[n] \quad i = 1, \ldots, 2 \tag{2.5}$$

where $x_i, y_i, z_i \in \mathbb{C}$ and the noise processes are i.i.d. over time with $z_i \sim \mathcal{CN}(0, N_0)$. By assuming that the channel from each transmitter to each receiver has a single tap we are restricting attention to the case of flat fading.

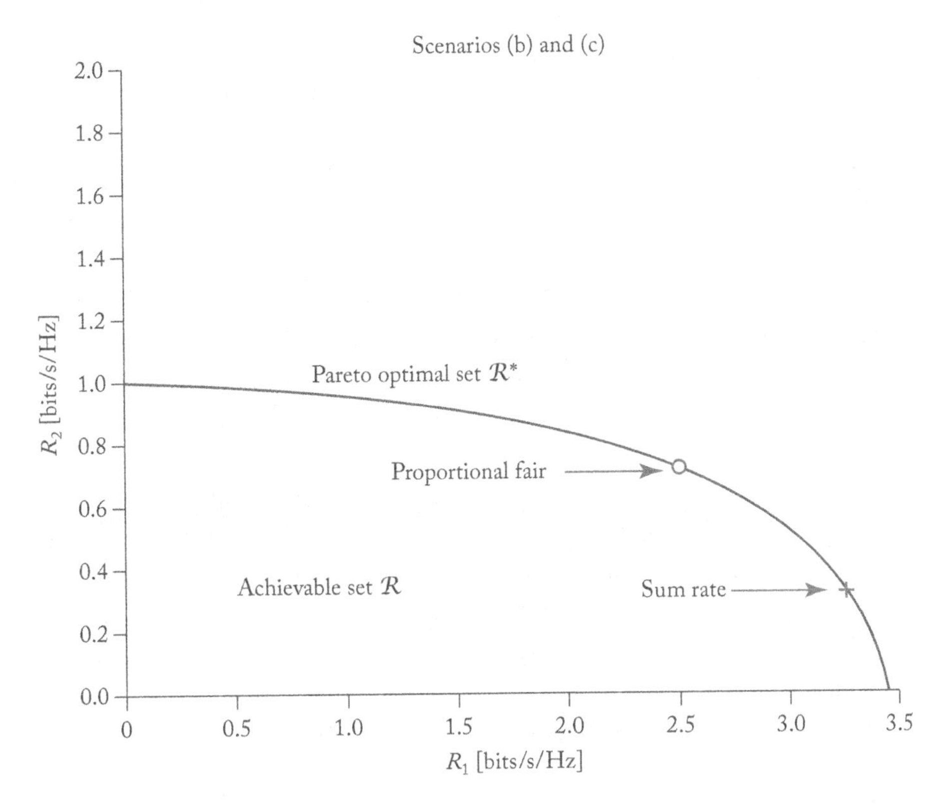

FIGURE 2.8: Achievable set $\mathcal{R}$ and Pareto efficient set $\mathcal{R}^*$ for scenarios (b) and (c).

trade off efficiency with fairness. Thus, we want to maximize some global utility function $U(R_1, R_2)$, that represents some fairness objective.

For concreteness we assign specific parameter values to each scenario in Figure 2.7. Without loss of generality we can assume in all cases that $c_{1,1} = c_{2,2} = 1$, $N_0 = 1$, and $W = 1$. In scenario (a) we choose $P_1 = P_2 = 10$, $c_{1,2} = 10$, and $c_{2,1} = 0.5$. For scenario (b) we set $P_1 = 10$, $P_2 = 1$, and $c_{1,2} = c_{2,1} = 1.1$. Finally, in (c) we set $P_1 = 10$, $P_2 = 1$, $c_{1,2} = 0.5$, and $c_{2,1} = 10$.

We assume that all the parameters are known to all the systems performing the optimization. In particular, we assume that the number of systems sharing the spectrum is common knowledge. Figure 2.8 shows the achievable set $\mathcal{R}$ and Pareto efficient set $\mathcal{R}^*$ (see Section 1.2) for scenarios (b) and (c).[4] For any utility function that is component-wise monotonically increasing in (R_1, R_2), the optimal rate allocation must occur in a point of the boundary $\mathcal{R}^*$. The choice of the utility function

4. The reason why for this specific choice of parameters in both scenarios we obtain the same sets will be explained later in this section.

strongly influence the fairness in the resulting allocations. For example when $U_{sum}(R_1, R_2) = R_1 + R_2$ the resulting optimal allocations are very unfair for system 2 ($R_2 \ll R_1$) (see Figure 2.8). (For scenario (a) this is the optimal allocation and gives equal rate to both systems.) A more fair allocation results from choosing the proportional fair metric $U_{PF}(R_1, R_2) = \log(R_1) + \log(R_2)$. We can see in Figure 2.8 how in scenarios (b) and (c) the use of the proportional fair metric results in a more fair allocation. Note that in scenario (a) the use of U_{PF} results in the same rates as when U_{sum} is used.

For more on the issue of fairness in this context see [Lan et al. 2010, Joe-Wong et al. 2012].

We still have to answer the question of how power is allocated by frequency for each of the two systems. Intuitively, when interference is very high, it is better to assign non-overlapping frequency bands. In [Etkin et al. 2007] it is shown that if $c_{i,j}c_{j,i} > c_{i,i}c_{j,j}$ then the Pareto efficient power allocations $p_i(f)$ and $p_j(f)$ are orthogonal, i.e., $p_i(f)p_j(f) = 0$ for all $f \in [0, W]$.

The condition $c_{i,j}c_{j,i} > c_{i,i}c_{j,j}$ means that for systems i and j, the product of the channel cross gains $c_{i,j}c_{j,i}$ is greater than the product of the channel direct gains $c_{i,i}c_{j,j}$. Note that the condition can be satisfied even if one of the cross gains is small, by having the other cross gain large enough. Also, note that the condition is independent of the power constraints $\{P_i, P_j\}$ and noise variance N_0. It is easy to check that this high interference condition is satisfied in our three examples due to our choice of parameters. In particular, if $c_{i,j}c_{j,i} > c_{i,i}c_{j,j}$ for any $i \neq j$, $j = 1, \ldots, M$, we can achieve any Pareto efficient rate vector with frequency division multiplexing (FDM).

Note that since the Pareto efficient rates are obtained with orthogonal allocations when $c_{1,2}c_{2,1} > 1$ (for direct gains equal to 1), the actual values of the cross gains $c_{1,2}$ and $c_{2,1}$ have no influence on the achievable region. This explains why scenarios (b) and (c) result in the same achievable region and optimal rates.

Note that for general channel conditions maximizing total data rate is a non-convex problem. In some cases it is reasonable to assume a kind of monotonicity which leads to globally optimal solutions [Qian et al. 2009].

2.5 MULTIPLE ACCESS

In the previous section we allocated spectrum resources by controlling transmitter power. Another method is to control access to the wireless channel.

Model 2.1 Consider N devices that share a wireless channel to send packets to each other, as shown in Figure 2.9. Each device is attached to a transceiver and an antenna. In today's technology, a transceiver cannot receive while it is transmitting, because the power of the transmitter overwhelms the receiver and makes it unable to hear another transmission. If transmissions of multiple

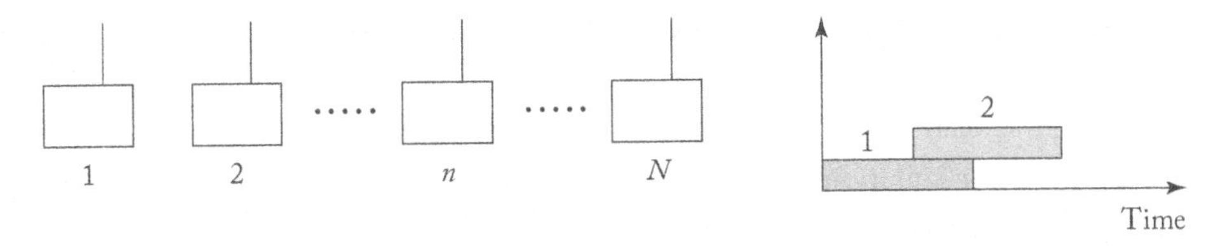

FIGURE 2.9: N devices share a wireless channel by transmitting at different times. Simultaneous transmissions such as 1 and 2 are corrupted and useless.

transmitters overlap in time, these signals add up at the receivers and they garble one another, which makes the receiver unable to decode correctly the transmissions. In such a situation, one says that the transmissions *collide*. The transmitters become aware of the collision when they do not get acknowledgements of the packets they sent. The problem is to regulate the transmissions by the different transmitters in order to reduce the frequency of collisions. The regulation scheme must be distributed and efficient. Moreover, it should adapt to the changing number of active transmitters and to the urgency or importance of the transmissions.

A simple scheme, called *time division multiplexing*, divides time into slots and allocates them in a round-robin order to the devices. When using this scheme, device i transmits in the time slots $i, N + i, 2N + i$, etc. However, this scheme is very inefficient if not all devices always have packets to transmit. Moreover, this scheme requires a good synchronization of the clocks and knowing how many devices are present. Myriads of other schemes have been devised to share a channel to improve the efficiency and the simplicity.

A widely used regulation mechanism is the *binary exponential back-off algorithm*. (See [Metcalfe 1973].) When using this scheme, a transmitter waits for the wireless channel to be idle, then waits for an additional random back-off time and then transmits if the channel is still idle. The transmitter chooses the random back-off time uniformly in $\{0, T, 2T, \ldots, (2^n - 1)T\}$ if the transmissions of the current packet have already collided n times. Here, T is some constant. The intuitive justification for this scheme is that if transmissions collide repeatedly, there are probably many active transmitters and they should randomize their back-off times over a larger set to reduce the chance of further collisions.

The performance of such a scheme can be estimated with an approximate model, when all the transmitters can collide with each other. The estimates of the achievable transmission rates by the transceivers that this approximation calculates seem to be reasonably accurate compared to simulations and they show that the scheme is not close to being the best possible.

In an alternative scheme, a transmitter uses random back-off times whose distribution is based on its backlog instead of the number of previous collisions. Remarkably, this modified scheme approaches the maximum possible total utility of the transmissions. However, this improved throughput is achieved at the cost of delays that may be rather long. The delay performance can be improved by using the delay of the packets to calculate the back off times. Also, this scheme works without modification in a situation where only subsets of the transmitters may collide. We explain the backlog-based scheme in the next section.

The intuition behind the algorithm is very simple: if the backlog in one device increases, that device should request the channel more urgently. Also, if the backlog decreases, so does the urgency of the requests. What is not obvious without analysis is how to adjust the distribution of the back-off times and how to control the arrivals in the queues to make the overall system fair.

Q-CSMA

Assume that any two transceivers collide with one another when their transmissions overlap. As a simplification, assume that the circuits of the transceiver are so fast that they can detect another transmission as soon as it starts. That is, if a device starts transmitting at time t, then any other device can hear that transmission right away and does not transmit, so that collisions do not happen for devices that choose different back-off times. Under this idealized situation, if transceivers choose independent back-off times that have a density, they never collide. The practical situation can also be analyzed, but is somewhat more complex.

Consider N transmitters. The goal is to design a scheme that maximizes the sum $\sum_i U_i(x_i)$ of the utilities of the transmission rates x_i of the transmitters. Here, $U_i(\cdot)$ is a concave increasing function. The scheme, as we will see, is such that transmitter i chooses an independent exponentially distributed back-off time with a mean value that decreases with its backlog q_i. That is, a transmitter becomes more impatient as its backlogs increases. Specifically, we assume that transmitter i requests the channel after an exponential time with rate $\exp\{\alpha q_i\}$, i.e., with mean $\exp\{-\alpha q_i\}$ where α is a positive constant. Moreover, transmitter i accepts new packets to transmit with the rate x_i that maximizes

$$U_i(x_i) - \beta x_i q_i,$$

where β is some positive constant. Thus, a transmitter reduces the rate at which it accepts new packets when its backlog increases. We call this scheme the *Q-CSMA* algorithm. It is a particular case of the situation we explored in Section 2.3.

To review how this scheme works, assume that the backlogs q_i are essentially constant for some relatively long period of time. Consider the Markov chain $\{z_t, t \geq 0\}$ where z_t indicates which transmitter is currently active. To make this process a Markov chain, we assume that the transmission times are all exponentially distributed with mean 1 (see Figure 2.10).

Thus, z_t takes values in $\{0, 1, \ldots, N\}$ and the non-zero transition rates are $q(0, i) = \exp\{\alpha q_i\}$ and $q(i, 0) = 1$. (It should be noted that, to make the channel request process independent of the

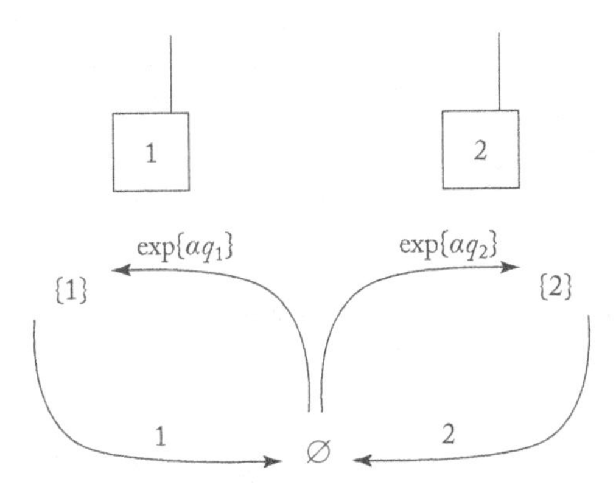

FIGURE 2.10: Two transmitter share a channel (top). The Q-CSMA algorithm corresponds to a Markov chain whose state is the set of active transmitters (bottom).

backlogs, a transmitter also requests the channel when its buffer is empty. Also, it keeps the channel for an exponential time with mean one in that case, which corresponds to sending a "dummy" packet.) The invariant distribution $\pi = \{\pi(i), i = 0, \ldots, N\}$ of this Markov chain can be seen to be given by

$$\pi(i) = A \exp\{\alpha q_i\}, i = 1, \ldots, N,$$

where $A = \pi(0)$ is the constant such that these probabilities add up to one. Indeed, one sees that

$$\pi(i)q(i,0) = \pi(0)q(0,i),$$

so that the Markov chain is time-reversible and π is its invariant distribution. As a verification, note that these identities are in fact the balance equations for state i since the only possible transitions out of state i and into state i are to and from state 0.

Thus, transmitter i gets to transmit a fraction $\pi(i)$ of the time, which means that its transmission rate x_i is equal to $\pi(i)$ since the average transmission time of a packet is equal to 1. Consider the optimization problem

$$\text{Maxmize } \sum_i U_i(x_i) - \beta \sum_i \pi(i) \log(\pi(i))$$
$$\text{subject to stability.} \tag{2.8}$$

To enforce stability, we introduce the drift of the function $V(\mathbf{q}) = \frac{1}{2}||\mathbf{q}||^2$, as we did in Section 2.3. That is, we consider the problem of maximizing

$$\sum_i U_i(x_i) - \beta \sum_i \pi(i) \log(\pi(i)) - \alpha \sum_i q(i)[x_i - \pi(i)]$$

since the drift of $(1/2)q_i^2$ is $q(i)[x_i - \pi(i)]$. The analysis of Section 2.3 then shows that the solution of the problem will correspond to the random access scheme that we discussed above.

Summing up, we find that the Q-CSMA algorithm solves the optimization problem 2.8. By making β very small, we see that this problem is essentially equivalent to maximizing the total utility of the transmission rates. The relaxation obtained by adding the correction terms to the sum of the utilities enables us to find a specific distribution that can be implemented by a randomized multiple access scheme. The analysis of the actual system in discrete time shows that the distance from the optimal performance depends on α.

Technically, one has to show that the Markov chain that describes the random access scheme converges fast enough compared to the changes in queue lengths. Indeed, our calculations above assume that the queue lengths remain constant long enough for the Markov chain to approach its invariant distribution. This analysis requires some bounds on the mixing (convergence) time of the Markov chain. All this works out, as shown in the reference.

Many variations of this algorithm are known. One can show that a simple variation applies to the case of store-and-forward ad hoc networks. Another variation achieves utility-optimal rates even in the practical situation where collisions are unavoidable because transmissions that start approximately at the same time collide. See [Jiang and Walrand 2010] for details.

2.6 STRATEGIC USERS

We observed in Section 2.2 that most resource allocation schemes are not strategyproof. Typically, these schemes assume that the users do not attempt to manipulate the mechanism to gain an advantage. In this section, it is shown that when users share a single link, the sum of the user utilities can decrease by up to 25% when the users are strategic. However, if they share paths in the network, the loss of efficiency can be very large.

Consider n users who share a link with capacity C. User i sends traffic at rate x_i and derives a utility $U_i(x_i)$, where $U_i(\cdot)$ is a concave non-negative increasing function with $U_i(0) = 0$.

SOCIAL OPTIMAL

Recall that to maximize $\sum_{i=1}^n U_i(x_i)$ subject to the capacity constraint $\sum_{i=1}^n x_i \leq C$, one forms the Lagrangian

$$L(\mathbf{x}, \lambda) = \sum_{i=1}^{n} U_i(x_i) - \lambda(\sum_{i=1}^{n} x_i - C),$$

where $\lambda > 0$ is a price per unit of excess rate. One then maximizes L over $\mathbf{x}$, which yields $\mathbf{x} = \mathbf{x}(\lambda)$, and one chooses λ so that the resulting rates satisfy the capacity constraint with equality. To maximize over $\mathbf{x}$, each user i chooses the value $x_i(\lambda)$ of x_i that maximizes his net utility

$$U_i(x_i) - \lambda x_i.$$

One then chooses λ so that $\sum_{i=1}^{n} x_i(\lambda) = C$.

Equivalently, one may consider that user i chooses to pay a total price $p_i = x_i\lambda$ for the link so as to get the rate $x_i = p_i/\lambda$. The corresponding net utility is

$$U_i\left(\frac{p_i}{\lambda}\right) - p_i. \tag{2.9}$$

With this formulation, user i chooses the value p_i of his payment, which depends on λ, that maximizes that expression. One then chooses λ so that

$$\frac{\sum_{i=1}^{n} p_i}{\lambda} = C. \tag{2.10}$$

This value of λ then determines p_i and the utility $U_i(p_i/\lambda)$ that we denote U_i^*. Define $U^* = \sum_{i=1}^{n} U_i^*$. Thus, U^* is the maximum total user welfare given the capacity constraint. We call U^* the *optimal social welfare*.

STRATEGIC USERS

In the analysis above, we assumed that the users are *price-takers*. That is, they consider that the price λ per unit rate in (2.9) is independent of their choice of p_i. However, we know from (2.10) that this is not the case. If the users are *strategic*, they anticipate the effect of their payment p_i on the unit price λ. That is, they optimize (2.9) where λ is given by (2.10). Thus, user i chooses the value p_i that maximizes

$$U_i\left(\frac{p_i C}{\sum_{j=1}^{n} p_j}\right) - p_i. \tag{2.11}$$

This maximization determines each p_i as a function of $\sum_{j \neq i} p_j$. This system of n equations in $(p_1, \ldots, p_n)$ admits one solution that is the *Nash equilibrium*[5] for the strategic users. This Nash equilibrium corresponds to values for the utilities that we denote by U_i^s. Let also $U^s = \sum_{i=1}^{n} U_i^s$, which is the user welfare when the users are strategic.

To measure the impact of selfishness, one defines the ratio POA, called the *price of anarchy*, as

$$POA = \frac{U^*}{U^s}.$$

5. A Nash equilibrium is a set of choices of the different users so that no single user can benefit by changing his choice unilaterally.

For instance, if $POA = 1.3$, then the total user welfare is divided by 1.3 when the users are strategic, compared to when they are price-takers. Thus, a large price of anarchy means that the mechanism is susceptible to user manipulation that deteriorates the user welfare. If that is the case, schemes that are robust to manipulation should probably be designed. Accordingly, it is important to estimate the POA to understand how robust the scheme is to user manipulation. Our discussion follows [Johari and Tsitsiklis 2009].

LINEAR CASE

As a concrete example, assume that $U_i(x) = a_i x$ for $i = 1, \ldots, n$ with $a_1 = 1$ and $a_i \leq 1$ for $i = 2, \ldots, n$. Let us also choose $C = 1$ to simplify the notation. In this case, the maximization of the total user welfare is achieved by allocating all the link rate $C = 1$ to user 1. Then,

$$U^* = \sum_{i=1}^{n} U_i^* = U_1(1) = 1.$$

To find U_i^s, we have to maximize (2.11), i.e.,

$$a_i \frac{p_i}{\sum_{j=1}^{n} p_j} - p_i.$$

Setting to zero the derivative of this expression with respect to p_i, we find

$$a_i \sum_{j \neq i} p_j = (p_1 + \cdots + p_n)^2,$$

i.e.,

$$a_i(x - p_i) = x^2 \quad \text{where } x = \sum_{j=1}^{n} p_j.$$

Thus,

$$p_i = x - \frac{x^2}{a_i}.$$

Summing over i, we find

$$x = nx - x^2 \sum_j a_j^{-1} \quad \text{or} \quad x = (n-1)(\sum_j a_j^{-1})^{-1}.$$

Hence,

$$U_i^s = \frac{a_i p_i}{x} = a_i - x$$

and

$$U^s = \sum_i a_i - nx = \sum_i a_i - n(n-1)\left(\sum_j a_j^{-1}\right)^{-1}.$$

Since $POA = 1/U^s$, to find the maximum possible value of POA, we minimize U^s over $(a_2, \ldots, a_n)$ subject to $a_i \in [0, 1]$ for $i = 2, \ldots, n$. By symmetry, this minimum occurs for $a_2 = \ldots = a_n =: y$. Substituting these values into U^s and recalling that $a_1 = 1$, we find

$$U^s = 1 + (n-1)y - \frac{n(n-1)}{1 + (n-1)y^{-1}} = 1 + (n-1)y - \frac{n(n-1)y}{n-1+y}.$$

Setting to zero the derivative with respect to y, we find that the minimizing value of y is $y = 1/2$. Substituting, we see that the minimum value of U^s is given by $(3n-1)/(4n-3)$. We conclude that

$$POA \leq \frac{4n-3}{3n-1} \leq \frac{4}{3}.$$

Also, that upper bound is tight since it is approached as $n \to \infty$.

GENERAL CASE

It turns out that this bound 4/3 on the price of anarchy for linear utilities is also an upper bound or arbitrary concave increasing utility functions that are equal to zero at zero. (This result is from [Johari and Tsitsiklis 2004].) To show this, we prove that the maximum POA occurs for linear utilities. The argument is that the value of POA that corresponds to given concave increasing utility functions $U_i(\cdot)$ with $U_i(0) = 0$ is not decreased by replacing the functions U_i by piecewise linear functions that have the same social optimum $\mathbf{p}^*$ and Nash equilibrium $\mathbf{p}$. Moreover, that POA is not decreased by replacing these piecewise linear functions by linear utilities.

First note that if $\mathbf{p}$ is a Nash equilibrium, each p_i achieves the maximum of (2.11), so that, again with $C = 1$, setting to zero the partial derivative of (2.11) with respect to p_i, we find

$$U_i'\left(\frac{p_i}{\sum_j p_j}\right) = \frac{(\sum_j p_j)^2}{\sum_{j \neq i} p_j} =: b_i, \quad \forall i,$$

i.e.,

$$U_i'(s_i) = b_i, \quad \forall i \text{ where } s_i := \frac{p_i}{\sum_j p_j}.$$

On the other hand, the value of U^* corresponds to $\mathbf{p}^*$ such that

$$U_i'\left(\frac{p_i^*}{\sum_j p_j^*}\right) = 1, \quad \forall i,$$

that is,

$$U_i'(o_i) = 1, \quad \forall i \text{ where } o_i := \frac{p_i^*}{\sum_j p_j^*}.$$

Second, assume that $\mathbf{p}$ is the Nash equilibrium and $\mathbf{p}^*$ the social optimal for the functions $U_i(\cdot)$. They are also the Nash equilibrium and the social optimal for any other functions, say $V_i(\cdot)$ such that

$$V_i'(s_i) = b_i \quad \text{and} \quad V_i'(o_i) = 1, \ \forall\, i.$$

Among all the concave increasing functions $V_i(\cdot)$ that satisfy these equalities and have a given value of $V_i^s = V_i(s_i)$, the one that has the largest value $V_i^* = V_i(o_i)$ is piecewise linear and has slope b_i between 0 and o_i and slope 1 to the right of o_i, as shown in Figure 2.11.

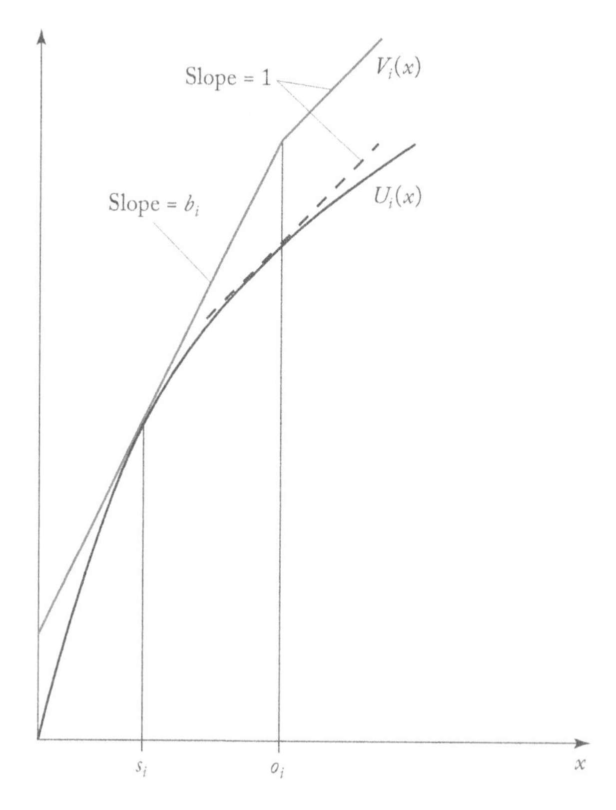

FIGURE 2.11: The functions $U_i(\cdot)$ and $V_i(\cdot)$.

For these functions V_i, the value of POA is given by the following expression:

$$POA = \frac{V^*}{V^s} = \frac{\sum_i V_i(o_i)}{\sum_i V_i(s_i)} = \frac{\sum_i(\alpha_i + b_i o_i)}{\sum_i(\alpha_i + b_i s_i)} \leq \frac{\sum_i b_i o_i}{\sum_i b_i s_i},$$

where the inequality comes from the fact that the α_i are non-negative and the ratio on the right is at least one. Now,

$$\frac{\sum_i b_i o_i}{\sum_i b_i s_i} \leq \frac{\max_j b_j}{\sum_i b_i s_i}$$

and the right-hand side is the price of anarchy when the utilities are $W_i(x) = b_i x$ for $i = 1, \ldots, n$, as we saw in the linear case.

Hence, the price of anarchy is bounded by 4/3.

SHARING MULTIPLE LINKS

A more realistic scenario than sharing a single link is when users have different paths through the network. In such a situation, representative of the Internet, the users are not aware of the path that their connections use. Accordingly, they cannot really strategize on the impact of their demand on the prices that links charge. We explore what happens when users try to be strategic.

As an example, assume that user 0 goes through links $1, 2, \ldots, n$ that have each a capacity $C = 1$ and that user i goes only through link i, for $i = 1, \ldots, n$. Assume that $U_i(x) = x$ for $i = 1, \ldots, n$ and $U_0(x) = ax$ where $a \leq n$. The social optimal corresponds to $x_0 = 0$ and $x_i = 1$ for $i = 1, \ldots, n$, so that $U^* = n$.

In the case of strategic users, first assume that user $i = 0, 1, \ldots, n$ pays p_i and gets a rate p_i/λ where λ is the price per unit rate. Thus, this model assumes that the network charges each user the same unit price λ. That is, in this first model, the *network* does not discriminate among users. The users choose the prices p_i they are willing to pay to maximize their net utility. By symmetry, one expects $p_1 = \ldots = p_n$. To meet the capacity constraints $x_0 + x_i \leq 1$, the network chooses λ such that

$$\frac{p_0}{\lambda} + \frac{p_i}{\lambda} = 1 = \frac{p_0 + p_1}{\lambda},$$

so that

$$\lambda = p_0 + p_1.$$

Accordingly, user 0 chooses p_0 to maximize

$$U_0\left(\frac{p_0}{\lambda}\right) - p_0 = \frac{ap_0}{p_0 + p_1} - p_0,$$

so that

$$ap_1 = (p_0 + p_1)^2.$$

Similarly, user 1 (and the users $2, \ldots, n$) choose p_1 to maximize

$$U_1\left(\frac{p_1}{\lambda}\right) - p_1 = \frac{p_1}{p_0 + p_1} - p_1,$$

so that

$$p_0 = (p_0 + p_1)^2.$$

Algebra shows that

$$p_0 = \frac{a^2}{(1+a)^2} \quad \text{and} \quad p_1 = \frac{a}{(1+a)^2}.$$

Hence,

$$U^s = U_0\left(\frac{p_0}{\lambda}\right) + nU_1\left(\frac{p_1}{\lambda}\right) = \frac{a^2+n}{a+1}.$$

Consequently,

$$POA = \frac{U^*}{U^s} = \frac{na+n}{a^2+n}.$$

Choosing $a = \sqrt{n}$, one finds

$$POA = \frac{1+\sqrt{n}}{2}.$$

Thus, $POA \to \infty$ as $n \to \infty$. What is happening in this example is that the network does not charge user 0 sufficiently for the n links he uses. As a consequence, that user's demand is excessive and reduces the utility of n other users.

A more suitable model is one where each link charges a price to each user, so that user 0 pays $n\lambda$ and user i pays λ per unit rate, for $i = 1, \ldots, n$. That is, this second model assumes that the *links* do not discriminate among users. Thus, if user 0 pays p_0 and every other user pays p_1, the network chooses the price λ so that

$$\frac{p_0}{n\lambda} + \frac{p_1}{\lambda} = 1,$$

i.e.,

$$p_0 + np_1 = n\lambda. \tag{2.12}$$

Now, assume that each user is unaware of his path through the network and makes the incorrect assumption that he uses m links and that there are on average k users using each link. Each user i pays a total price per unit rate equal to p_i, so that user i assumes that he pays each link j at rate p_i/m, since he thinks he uses m links. Thus, user i assumes he gets a rate

$$\frac{p_i/m}{(p_i/m) + (q_i/m)} = \frac{p_i}{p_i + q_i}$$

on each link, where q_i/m is the amount that the other users pay on each link. Trying to be strategic, user i then chooses p_i to maximize

$$U_i\left(\frac{p_i}{p_i + q_i}\right) - p_i.$$

Thus, user 0 chooses p_0 so that

$$aq_0 = (p_0 + q_0)^2.$$

Similarly, each other user i chooses p_i so that

$$q_i = (p_i + q_i)^2.$$

By symmetry, each user i assumes that $q_i = (k-1)p_i$ since he thinks that k users identical to himself share each link. Thus, user 0 calculates

$$a(k-1)p_0 = (p_0 + (k-1)p_0)^2,$$

so that

$$p_0 = \frac{a(k-1)}{k^2}.$$

Similarly, each other user i calculates

$$(k-1)p_i = (p_i + (k-1)p_i)^2,$$

so that

$$p_i = \frac{k-1}{k^2}.$$

The resulting sum of utilities is then U^s where

$$U^s = a\frac{p_0}{\lambda} + n\frac{p_1}{\lambda} = \frac{ap_0 + np_1}{p_0 + np_1}$$

$$= \frac{a^2(k-1) + n(k-1)}{a(k-1) + n(k-1)} = \frac{a^2 + n}{a + n}.$$

Since $U^* = n$, we find that the price of anarchy POA is given by

$$POA = \frac{an + n^2}{a^2 + n}.$$

As before, we see that this price of anarchy can be quite large. For instance, if $n = 10$ and $a = 2$, we find $POA \approx 8.6$. In this example, the price of anarchy arises because the users are unaware of their path through the network, even though the networks knows those paths and charges the users accordingly. That is, users try to be strategic but make incorrect assumptions about their externality.

SPECTRUM SHARING WITH STRATEGIC USERS

In this section we extend the model of Section 2.4 by assuming that the users are strategic, i.e., they choose strategies to maximize their own utilities without regard to the utilities of others. This game is known as the Gaussian Interference Game (GIG). A strategy s_i for a given user is the choice of power allocation $p_i(f)$. The players play simultaneously, and know the utility functions of the other players. Thus, this is a game of perfect information.

It turns out that a direct consequence of the flat-fading and white noise assumption is:

The set of frequency-flat allocations $p_i(f) = P_i/W$, $f \in [0, W]$ for $i = 1, \ldots, M$ is a Nash Equilibrium of the GIG.

This means that the best possible strategy for a given system is to spread its available power over the total bandwidth whenever all the interfering systems are spreading their signals. Intuitively, the best response of a system to the choices of the other users is to waterfill the available power over the noise+interference seen. When all the other systems use flat allocation, the waterfilling power allocation is flat, and it follows that flat allocations are best responses to each other.

In many cases, the set of rates that results from the full-spread Nash equilibrium is not Pareto efficient (i.e., is not in $\mathcal{R}^*$) so there may be a significant performance loss if the M systems operate in this point due to lack of cooperation. And in many cases this inefficient outcome is the only possible outcome of the game [Etkin et al. 2007]. In fact, the loss of efficiency (over for example, having the various users orthogonalize) can be made arbitrary large.

As in many such scenarios, a more realistic way to think about the situation is that the systems are likely to be in place for a long time and therefore will be playing a "repeated game" as opposed to a "single shot" one. Thus, it is reasonable to model the scenario as a repeated game where systems play multiple rounds, remembering the past experience in the choice of the power allocation in the next round. We will consider an infinite horizon repeated game, where the GIG is repeated forever. The utility of each player is defined by

$$U_i = (1 - \delta) \sum_{t=0}^{\infty} \delta^t R_i(t), \tag{2.13}$$

where $R_i(t)$ is the utility of user i in the stage game at time t, and $\delta \in (0, 1)$ is a discount factor that accounts for the delay sensitivity of the systems. At the end of each stage, all the players can observe the outcome of the stage-game and can use the complete history of play to decide on the future action. A strategy in the repeated game is a complete plan of action, that defines what the player will do in every possible contingency in which he may need to act.

One property of this repeated game is that sequences of strategy profiles that form a Nash Equilibrium (N.E.) in the stage game, form a N.E. in the dynamic game[6]. Furthermore, the dynamic game allows for a much richer set of N.E. This is an advantage from the point of view of policy making or standardization. The systems can agree through a standardization process to operate in any N.E. of the dynamic game. Having many equilibrium points to choose from gives more flexibility in obtaining a fair and efficient resource allocation. A natural question that arises is what set of rates can be supported as a N.E. of the repeated game. The following result is a general version of which is due to Friedman [Friedman 1971, Friedman 1977] and gives a sufficient condition for the rate vector $(R_1, \ldots, R_M)$ to be achievable as the resulting utilities in a N.E. of the repeated game.

Let R_i^{FS} be the rate of system i when all the systems spread their power over the bandwidth W, i.e., the rate obtained in the full-spread N.E. There exists a sub-game perfect N.E.[7] of the dynamic Gaussian Interference Game with utilities $(U_1, \ldots, U_M) = (R_1, \ldots, R_M)$ whenever $(R_1, \ldots, R_M) \in \mathcal{R}$ and $R_i > R_i^{FS}$ for $i = 1, \ldots, M$ for a discount factor δ sufficiently close to 1. For a proof, see [Etkin et al. 2007].

Let $\{p_i(f)\}_{i=1}^M$ be the power allocations that result in the rate vector $(R_1, \ldots, R_M)$ (which always exist since $(R_1, \ldots, R_M) \in \mathcal{R}$).

The strategy that each system follows to obtain the rate vector $(R_1, \ldots, R_M)$ in the result is the following trigger strategy:

- at $t = 1$: use power allocation $p_i(f)$; and
- at $t = t_0$: if at time $t = t_0 - 1$ every user $j \in \{1, \ldots, M\}$ used the power allocation $p_j(f)$ then use $p_i(f)$. Otherwise spread the power over the total band, i.e., use the power allocation P_i/W for $f \in [0, W]$

6. For the reader familiar with game theory, these equilibria are in fact sub-game perfect Nash equilibria.

7. The sub-game perfect N.E. is a refined and stronger version of the N.E. concept defined before. It guarantees that the N.E. does not arise due to unbelievable threats.

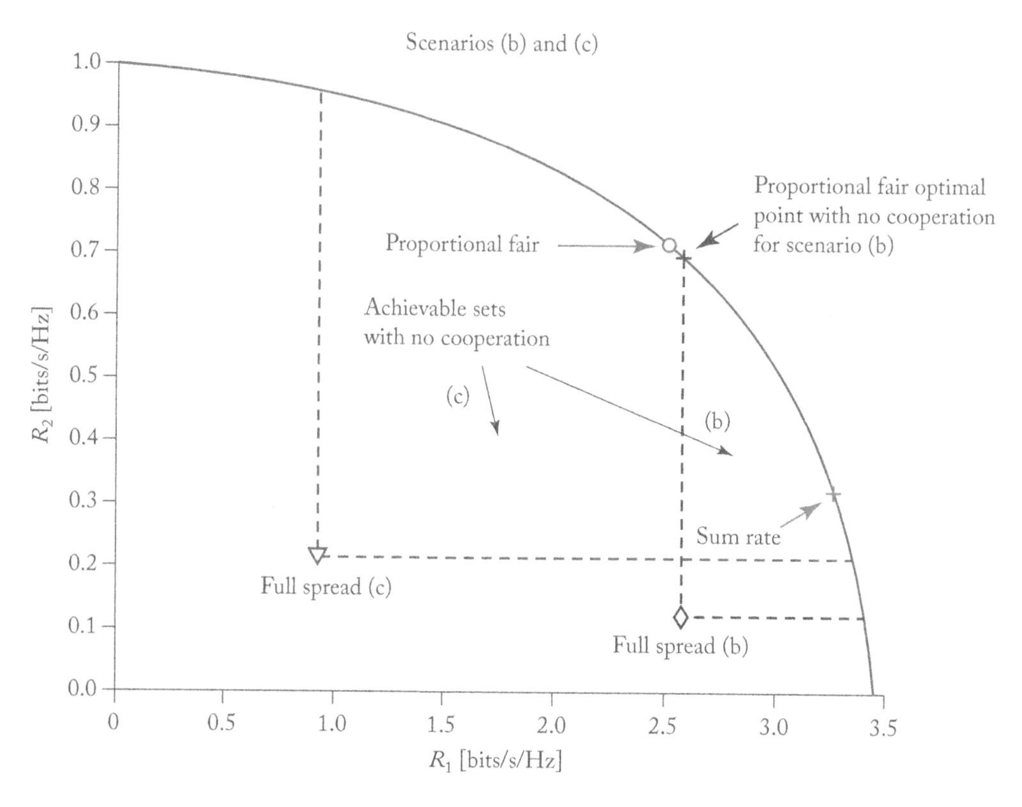

FIGURE 2.12: Achievable rates with no cooperation for scenarios (b) and (c).

The idea behind this strategy is to "cooperate" by using the required power allocation as long as all the other systems cooperated in the previous stages. As soon as at least one system deviates from the "good" behavior, a punishment is triggered where all the other systems spread their powers forever. Since the rates obtained by the systems once the punishment is triggered are lower than those obtained with cooperation, it is in the system's own interest to cooperate. Friedman's analysis shows that if δ is not too small, the above set of strategies forms a sub-game perfect Nash equilibrium. The sub-game perfection property of the Nash equilibrium guarantees that each system will indeed apply the punishment once the punishing situation arises. This property makes the threats believable.

Referring to Figure 2.12 we see that in scenario (b) of Figure 2.7 the optimal sum rate point lies within the achievable region in the non-cooperative setting. However, the optimal proportional fair point lies outside of this set and cannot be supported without cooperation. The best that one can do in the non-cooperative setting is to operate in the point indicated in the figure. In scenario (c) both the optimal sum rate and optimal proportional fair rates are achievable in the non-cooperative setting. Note that while in the cooperative case the specific values of the cross gains had no influence on the achievable region (as long as the strong interference condition is satisfied) this is not true in the non-cooperative setting. This is because large cross gains enable the systems to apply punishments, and hence achieve a good Nash equilibrium through believable threats. In scenario (c) the large value of $c_{2,1}$ allows system 2 to punish system 1 whenever it departs from the proportional fair allocation.

2.7 SUMMARY AND REFERENCES

The last two decades have seen a considerable amount of research on distributed algorithms that underly networking protocols. That research has resulted in systematic methods to design protocols with provable throughput and delay properties. This chapter explains some of the main insights.

The idea that a distributed transport protocol such as TCP might solve a network utility maximization problem was recognized by Kelly [Kelly 1997]. Mo [Mo and Walrand 2000] proposed the idea of α-fairness to achieve a trade-off between maximizing the sum or the minimum of the user utilities. Backpressure protocols are justified with a slightly different formulation of the utility maximization problem. In this formulation, the routers store packets of different flows in distinct queues and they decide which flows to serve at any given time. This variation enables to also consider routing as a control variables. The result is a network where the routers make scheduling and routing decisions on a packet-by-packet basis based on real-time information from their downstream routers. The resulting protocols are more responsive to changes in link characteristics such as failures or fading than traditional layered protocols that enforce a time-scale separation of the different protocols. Such ideas may one day have an impact on the design of wireless systems or data center networks. See [Huang and Walrand 2013] for a concise presentation of those ideas. The references [Shakkottai and Srikant 2008] and [Srikant and Ying 2014] provides a comprehensive coverage of the network utility maximization.

Tassiulas and Ephremides [Tassiulas and Ephremides 1992] introduced the method of minimizing the drift of a Lyapunov function to stabilize a network. Neely et al. [Neely et al. 2005] combined that idea with the problem of maximizing the utility. For a detailed presentation, see [Neely 2010]. The effect of strategic users is mitigated in a repeated version of the game, as we explained for a wireless system; see [Etkin et al. 2007] for details. Jiang [Jiang and Walrand 2010] introduced randomized matching algorithms for wireless networks. The distributed algorithms assume that the users are "price-takers." Johari and Tsitsiklis [Johari and Tsitsiklis 2004] analyzed the price of anarchy caused by strategic users. To avoid such a price of anarchy, one can use a VCG pricing, as was explored by Yang and Hajek [Yang and Hajek 200].

CHAPTER 3
Auctions

When a number of users compete for resources, one has to decide which users get what resources. If the goal is to maximize the usefulness of the resources, one has to elicit information about how much users value the resources. That is, one has to design a "mechanism" to which users respond by making requests for resources and the mechanism determines the resource allocation and the prices. Auctions are a basic class of such mechanisms and they are used extensively in position auctions by search engines and in spectrum auctions. Auctions generate much of the funding for the Internet.

In many cases, one considers a single item. Auctions elicit preferences through bids. The agents, or bidders, place bids and an auctioneer, or seller, awards the resource to one of the bidders for a price that depends on the bids.

One considers two possible goals of the auction. Either the auction is designed to award the resource to the agent who values it the most. In this case, we say that the auction is *efficient* because it maximizes the user welfare. Or the auction is designed to maximize the revenue to the seller. Such an auction is said to be *optimal* (for the seller). The auctions are almost always designed to be *individually rational*, so that agents never pay more than their valuation. When the valuations of the agents are random, these goals are based on the expected values of the revenue or net utility for the agents.

The results in this section are somewhat surprising. For example, we will find it is frequently beneficial for the seller not to award the item to the highest bidder, and to sometimes choose to award it to no one even when the highest bid exceeds the seller's own valuation of the object. It is important to keep in mind that the seller is trying to maximize his revenue in expectation and not for every instance of the auction, and also that each bidder knows how the valuations of the other bidders are probabilistically distributed. We start with a simple example.

3.1 PRELIMINARIES

Suppose there are four bidders—1, 2, 3, 4—and one item. Bidder i has a private valuation $U(i)$ for the item. We want to allocate the item to the bidder with the largest valuation. Thus, we want to design an efficient auction.

FIRST PRICE AUCTION

One simple scheme is the *first-price sealed-bid auction* where the bidders put their bid in a sealed envelope that they give to the auctioneer and the item goes to the highest bidder for the price she bid. This is the way most houses are sold in California. This auction is also called a *Dutch auction.*

There is an obvious problem with this scheme: the bidders may shade their bid, i.e., bid less than their actual valuation, to attempt to pay less for the item. As a result, it is not even clear that the item will go to the bidder who values it the most.

How should the bidders shade their bid? Consider the case where the U_i are i.i.d. It can be shown that bidder i should calculate and bid $\phi(U_i)$ where

$$\phi(U_i) = E[Z_i|Z_i < U_i] \quad \text{where } Z_i = \max_{j \neq i} U_j.$$

For instance, if the U_i are uniformly distributed in $[0, 1]$, then $\phi(u) = 3u/4$. Thus, in this case, the auction is efficient.

In this auction, the expected revenue is the expected value of the second largest valuation, which is 0.6, whereas the expected valuation of the bidder who gets the item is 0.8.

Note that the bidders could all collude and pretend that their valuations are in fact $0.01 \times U_i$. However, the above strategy maximizes the expected net utility of each bidder, assuming no collusion. More precisely, this strategy is a Bayesian Nash equilibrium. [1]

SECOND PRICE AUCTION

Another scheme is to start with a low ask price and increase it slowly until only one bidder is willing to buy the item. This is how houses are sold in a public auction in Belgium. This is also how many art auctions are run. This auction is also called a *Vickrey auction* [Vickrey 1961] because Vickrey proved that it is strategyproof. The auction is also called an *English auction.*

In this case, the bidder with the maximum valuation gets the item and pays a price equal to the next largest valuation. That is, if user i gets the item, he pays Z_i. Thus, an equivalent implementation is a *second-price sealed-bid auction.*

Note that the expected revenue of the auction is the expected value of the second valuation. This is the same expected revenue as in the first price auction. This fact is a particular case of the *revenue equivalence theorem* that states that a large class of auctions have the same revenue. We explain that result in Section 3.5.

1. A Bayesian Nash equilibrium is such that no agent can increase his expected payoff by changing his strategy.

Moreover, the bidders do not have any incentive to lie about their valuation since what they pay does not depend on their actual bid. In fact, this scheme is a particular case of a VCG mechanism that we discuss in Section 3.2.

In this example, all the bidders could still collude.

MAXIMIZING REVENUE

The first and second price auctions yield the same expected revenue. It turns out that a different auction achieves a larger revenue. If the U_i are i.i.d. and uniformly distributed in $[0, 1]$, this alternative auction is an ascending auction with a starting price of $1/2$, called *reserve price*. This auction is, in fact, optimal.

This auction proceeds as follows. The auctioneer opens the biddings with the reserve price $1/2$. If no agent bids at that price, the auction stops and the item is not sold. Otherwise, if only one agent bids, he gets the item at the reserve price. If the two agents bid, the auctioneer increases the price slowly until only one agent still bids. The auction then stops and the remaining agent gets the item at that price.

For instance, if there are two bidders, with probability $1/4$, the two valuations are below the reserve price and the item is not sold. With probability $1/2$, one valuation is uniform in $[1/2, 1]$ and the other is uniform in $[0, 1/2]$, so that the item is sold at the reserve price $1/2$ (instead of at the second valuation as in a second price auction). With probability $1/4$, both valuations are larger than $1/2$ and the expected value of the smallest one is $1/2 + (1/3)(1/2) = 2/3$. Thus, the expected revenue of the auction is $(1/2)(1/2) + (1/4)(2/3) = 5/12$. This is larger than the expected value of the second price auction, which is $1/3$. We discuss this auction and its optimality in Section 3.3.

3.2 VCG AUCTION

The VCG auction, named after its designers Vickrey, Clark, and Grove, is incentive compatible: it is a dominant strategy for every user to be truthful, i.e., to declare his true valuation. Also, the allocation is efficient: it maximizes the sum of the utilities of the agents.

THE AUCTION

Consider a set $\mathcal{S}$ of items and N agents. Each agent $i = 1, \ldots, N$ has a utility $U_i(S)$ for every subset S of items. The VCG auction proceeds as follows. Each agent i declares a utility $V_i(S)$ for every $S \subset \mathcal{S}$. The agents do not have to be truthful. The auctioneer then calculates the allocation $\{S_1^*, \ldots, S_N^*\}$ of items to the N agents that maximizes the total declared utility of the agents. That is, the allocation maximizes

$$\sum_{i=1}^{N} V_i(S_i) \tag{3.1}$$

over all the possible collections $\{S_1, \ldots, S_N\}$ of (pairwise) disjoint subsets of $\mathcal{S}$. Moreover, for every agent i, the auctioneer calculates the reduction p_i in utility of the other agents caused by agent i. That is,

$$p_i = \max\{\sum_{j \neq i} V_j(S_j) \mid S_j \text{ disjoint}\} - \sum_{j \neq i} V_j(S_j^*). \tag{3.2}$$

Each agent i then pays p_i for his allocation S_i^*. One says that p_i is the *externality* of agent i and that the VCG auction *internalizes the externality* by making each agent pay for his externality.

To see why it is dominant for every agent i to be truthful, we first compute the net utility of agent i. It is

$$U_i(S_i^*) - p_i = U_i(S_i^*) + \sum_{j \neq i} V_j(S_j^*) - A_i,$$

where A_i denotes the first term in the expression (3.2) for p_i. Second, we note that A_i does not depend on the utility that agent i declares. Third, we recall that the auctioneer chooses the allocations $\{S_1^*, \ldots, S_N^*\}$ that maximize (3.1). Thus, if agent i declares a utility $V_i(\cdot)$, the allocation maximizes $V_i(S_i) + \sum_{j \neq i} V_j(S_j)$ but does not maximize $U_i(S_i) + \sum_{j \neq i} V_j(S_j)$, so that the net utility of user i, which is equal to $U_i(S_i) + \sum_{j \neq i} V_j(S_j) - A_i$, is maximized only if he declares his true utility $U_i(\cdot)$.

Since the agents declare their true utility, the auctioneer chooses the allocation that maximizes the sum of the utilities, so that the auction is efficient.

The VCG auction is complex. It requires every agent specifying his utility for all the possible subsets of items. Also, the auctioneer must solve $N + 1$ optimization problems: one for the allocation and one for each price. The optimization problems are generally hard unless the problem has more structure.

For instance, the second price auction is a VCG auction. Indeed, if bidder 1 has the highest bid, he gets the item and deprives the bidder with the second highest bid of the item. Hence, the winner reduces the utility of the other bidders by the amount of the second bid, which is precisely his payment.

Given the importance of the VCG mechanism, let us consider a few more examples:

Example 3.1 (Computing Minimum Cost Paths) Consider an inter-domain routing problem where we wish to establish a path between two nodes s and t in a graph with a set $\{1, \ldots, M\}$ of edges. Say that the edges are owned by different providers and that the provider of edge e faces a cost u_e for carrying traffic over the edge. The network operator runs a VCG auction that incentivizes the providers to declare their true costs.

The total welfare of the user of the path and the providers is the value of the service minus the cost of providing it. Thus, if the providers declare costs v_e for the edges, the network operator chooses the least-cost path from s to t. Say that this path uses the edge e and costs $v_e + w_e$. Thus, the sum of the costs to the providers other than e is equal to $-w_e$. Assume also that if edge e were not available, the cheapest path would cost $r_e > v_e + w_e$. That is, the sum of the costs to the providers other than e would be $-r_e$ if e were not present. Accordingly, the VCG price p_e for provider e is the reduction in welfare of the agents other than e when e is present, i.e.,

$$p_e = -r_e - (-w_e) = w_e - r_e.$$

Equivalently, provider e gets paid $r_e - w_e$. His net utility is $r_e - w_e - u_e$ since his actual cost is u_e. Thus, the net utility of provider e is

$$(r_e - w_e - u_e)1\{r_e > v_e + w_e\}$$

since edge e is used only if the cost with e is less than without. We can see that this net utility is maximized over v_e by choosing $v_e = u_e$, which once again verifies the incentive-compatibility of VCG.

As a simple example, assume that s and t are connected by M edges $\{1, \ldots, M\}$ in parallel. Assume the providers declare the costs $(v_1, \ldots, v_M)$ with $v_1 < v_2 < \cdots < v_M$. Thus, if the actual costs are $u_1 < u_2 < \cdots < u_M$, the network operator pays u_2 to the provider of edge 1 even though his cost is only u_1.

Figure 3.1 illustrates the VCG payments for a specific graph. Notice that in the case that the applications tell the truth, the network is overpaying. This can be viewed as the price paid for keeping the agents honest. See [Archer and Tardos 2007] for more details.

The example illustrates that even when an agent i tells the truth the payment may still be non-zero. To make this point more dramatically, suppose edge d in Figure 3.1 is broken up into n parts so that there is a path $s, d_1, d_2, \ldots, d_n, e, f$, and let the cost of the edges be $d_i = 2/n$. Then the cheapest path still costs 5, and each edge d_i must be paid $6 - (5 - \frac{2}{n})$, which for large n is 1. Thus, when everyone tells the truth, the edge owners are paid a total of $O(n)$ so that the network can save one unit of cost for the path (from 6 to 5)!

Of course, similar examples can be constructed in which the network extracts over-payment from the agents. Thus, VCG mechanisms are not *budget balanced*. This drawback can be quite severe, since the goal is to maximize utility, but we can end up doing considerably worse because of the payments.

It is natural to ask if one can actually design mechanisms which are *budget balanced*, i.e., the overall payment to the system is zero. It turns out that it is impossible to design efficient allocation

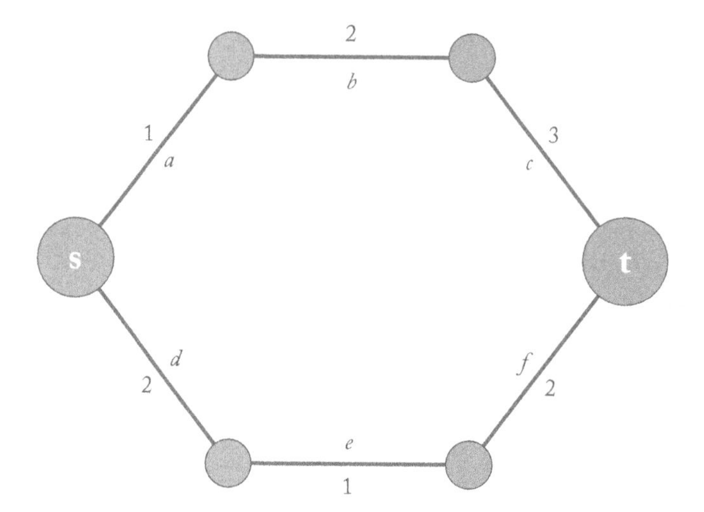

FIGURE 3.1: When all of the agents report their costs truthfully, the cheapest path uses the edges d, e and f and has a value of 5. Payments are 0 for edges a, b and c since are not part of the optimum, but are 3, 2 and 3 for d, e and f. Thus, while the welfare maximizing utility is 5 units, the edge owners have to be paid 8 units to achieve that cost.

schemes which allow arbitrary quasi-linear utility functions and are budget balanced [Green and Laffont 1977]. For much more on this subject see a standard reference on Microeconomics such as [Mas-Colell et al. 1995]. In Section 5.2, we explore a set of mechanisms that *is* budget balanced but not necessarily efficient.

Example 3.2 (Multicast Tree) Suppose we are given a tree that spans the nodes corresponding to the users in N and each edge has a weight that corresponds to the cost of building it. The cost of a subtree that spans $S \subseteq N$, $C(S)$, is the sum of the edge costs of the tree. We wish to allocate the costs of providing the service to users. Each user i values the multicast service at v_i dollars. Our goal is then to find the set S that maximizes the total welfare $W(S)$ where

$$W(S) = \sum_{i \in S} v_i - C(S).$$

Each user reports his value v_i. The problem is solved based on the reported values. Let U^{-i} be the maximum welfare when i is not present. Also, for $i \in S^*$, let

$$W^{-i} = \sum_{j \in S^* \setminus \{i\}} v_i - C(S^*)$$

and $W^{-i} = 0$ if $i \notin S^*$. The VCG payment from each node i is then p_i where

$$p_i = U^{-i} - W^{-i} = W(N \setminus \{i\}) + v_i - W(N).$$

Example 3.3 (Combination of Resources) Suppose the resources being assigned are licenses to transmit over a specific band of spectrum over different regions of the country, and the bidders are operators of communications services. The utility of a given bidder is best expressed as a a function of a subset of these licenses. For example, subsets that consist of licenses in contiguous regions may be more valuable than subsets that consist of licenses to fragmented, non-contiguous regions. The license granting body wishes to assign licenses to maximize the sum of the operator utilities.

Let L be the set of licenses so that for each $S \subset L$, $U_i(S)$ is the utility to operator i if he were granted S. Let $\mathcal{P}$ be the set of all partitions of S into N sets, where N is the number of operators. Under a partition, $\alpha \in \mathcal{P}$, let α_i be the licenses assigned to operator i for $i = 1, 2, .., N$.

Then the payment from operator i is

$$p_i = \max_{\alpha \in \mathcal{P}} \sum_{j \neq i} U_j(\alpha_j) - \sum_{j \neq i} U_j(\alpha_j^*),$$

where $\alpha^* = (\alpha_1^*, \ldots, \alpha_N^*)$ is the partition that maximizes the operator utility, based on the utility functions reported by the bidders.

Clearly, this scheme is impractical since there are $2^{|L|} - 1$ subsets of L.

This is an example of a combinatorial auction that has special relevance to the problem of managing spectrum. We will examine more practical schemes for managing this process in Section 3.4.

VCG AUCTION OF IDENTICAL ITEMS

Assume there are N identical items and M agents. Agent i has a utility $U_i(n)$ for $n = 1, 2, \ldots, N$ items. We assume that the functions $U_i(\cdot)$ are non-decreasing and concave. That is, the marginal utility $u_i(n) = U_i(n+1) - U_i(n)$ for agent i getting one more item when he already has n items is nonnegative and non increasing.

Ausubel [Ausubel 2004] derived an ascending auction that generalizes the ascending implementation of a second price auction for a single item and implements the VCG auction. The Ausubel auction is as follows. Let $D_i(p)$ be the demand for items by agents other than agent i when the price is $p + \epsilon$. The auctioneer increases the price slowly from $p = 0$. Every time $(M - D_i(p))^+$ increases, agent i gets an item at price p.

To understand this auction, assume that $(M - D_i(p))^+$ just increased by 1 when the price reached the value p. This means that some agent other than i had a utility p for one more item.

Thus, p can be viewed as the externality caused by agent i getting one more item instead of some other agent.

3.3 OPTIMAL AUCTION

The VCG auction is efficient, but it does not maximize the revenue of the auction, as we saw in the case of agents with i.i.d. valuations uniform in $[0, 1]$. We explain the revenue maximizing auction due to Myerson [Myerson 1981].

MYERSON AUCTION

An auction is said to be *interim incentive compatible* if it is a dominant strategy, in expectation, for agents whose type is known to the auctioneer, to reveal their true valuation. An auction is *interim individually rational* if no agent can get a negative expected surplus by participating in the auction. Finally, an auction is said to be *optimal* if it maximizes the auctioneer's revenue among all incentive compatible and individually rational auctions.

There is a single item and N agents. The agents have independent valuations for the item. The valuation of agent i has a positive density $f_i(x)$ on $[a_i, b_i]$ that is zero elsewhere and we denote by $F_i(x)$ the cumulative distribution function. The number of agents and the distributions $F_i(x)$ are known to the auctioneer and to all the agents.

We assume that

$$c_i(x) := x - \frac{1 - F_i(x)}{f_i(x)} \tag{3.3}$$

is non-decreasing on $[a_i, b_i]$. This assumption holds for distributions with non-decreasing hazard rates, such as uniform, exponential, and Gaussian. (The case when $c_i(x)$ is not increasing in x is also analyzed in [Myerson 1981].)

The Myerson auction allocates the item to the agent with the maximum *virtual valuation* $c_i(U_i)$, provided that it is positive. Also, if agent i gets the item, his payment p_i is the minimum value of x that makes him win the auction.

Note that this auction is based on the true valuations of the agents. How does the auctioneer know these valuations? What if the agents misrepresent them? The key idea here is that the auctioneer can be viewed as a neutral intermediary to whom the agents might as well reveal their true valuation. Indeed, imagine that the best strategies for the agents is to declare valuations B_i instead of U_i. Agent i computes this best strategy as a function of the true valuation U_i, say $B_i = \phi_i(U_i)$. The auctioneer then gets these bids B_i and decides the outcome of the auction as some function $\psi(B_1, \dots, B_N)$. An equivalent implementation of this mechanism is that the agents reveal their true valuations U_i to the auctioneer who then calculates $\psi(\phi_1(U_1), \dots, \phi_n(U_n))$. In other words, any mechanism that the auctioneer can implements with the B_i can also be implemented with the U_i. That is, one may assume that the agents reveal their true valuations. This observation is Myerson's *revelation principle* [Myerson 1981].

The following result is proved in Section 3.5.

Theorem 3.1 The Myerson auction is optimal.

Let us look at simple examples. Say that the valuation of agent i is uniform on $[a_i, b_i]$. Then $c_i(x) = 2x - b_i$. Thus, the item goes to the agent with the maximum value of $2U_i - b_i$, provided that this maximum value is positive.

First, assume that $a_i = 0$ and $b_i = 1$ for all $i = 1, \ldots, N$. Then the item goes to the agent with the largest value of $2U_i - 1$, provided that it is positive. Thus, it goes to the agent with the highest valuation U_i provided that it is larger than 1/2. Say that agent i wins the auction. He pays the minimum value of x such that $2x - 1 \geq 0$ and $2x - 1 \geq 2U_j - 1$ for all $j \neq i$. This payment is the maximum of 1/2 and the second highest valuation. Hence, this auction is an ascending auction with a reserve price equal to 1/2.

Second, say that there are two bidders, one with $[a_1, b_1] = [0, 10]$ and the other with $[a_2, b_2] = [0, 20]$. Assume that $U_1 = 9$ and $U_2 = 13$. Then,

$$c_1(9) = 2 \times 9 - 10 = 8 \quad \text{and} \quad c_2(13) = 2 \times 13 - 20 = 6.$$

Thus, the item goes to bidder 1 and he pays the minimum value of x so that

$$2x - 10 \geq 2 \times 13 - 20,$$

i.e., he pays 8. Thus, the item does not go to the user with the highest valuation. This auction favors the "poorest" bidder.

M Identical Items and N Agents

Multiple item auctions are generally complicated, as we will discuss later. There is one easy case, however: when each agent wants only one item.

There are M identical items and N agents. Each agent wants only one item. The agents have independent valuations $(U_1, \ldots, U_N)$ for one item. The valuation U_i are distributed as in the previous section and we again assume that $c_i(x)$ is non-decreasing on $[a_i, b_i]$ for each i.

The revenue-maximizing auction is to assign the items to the agents with the maximum and nonnegative virtual valuations $c_i(U_i)$. The proof is exactly the same as in the case of a single item (see Section 3.5).

For instance, if the valuations are uniform in [0, 1], the auctioneer ranks the U_i in decreasing order and assigns the items in decreasing order of the U_i as long as they are larger than 1/2.

If valuation i is uniform in $[a_i, b_i]$, then $c_i(v) = 2v - b_i$. The auctioneer orders the agents in decreasing values of $2U_i - b_i$ and assigns the M items in that order as long as the values are positive.

2 Items and 1 Agent

As we saw, the Myerson auction is optimal when each agent wants only one item. Otherwise, the situation gets very complicated. As an example, assume that there are two items and one agent. The agent has a random valuation(X_1, X_2) of the two items and the valuation of the pair is $X_1 + X_2$. The seller knows the distribution of that pair and the agent knows its realization.

For instance, say that X_1 and X_2 are i.i.d. with $P(X_1 = 10) = P(X_1 = 22) = 1/2$. What prices should the seller select for the items to maximize his revenue? If each item is offered for sale at price 10, the expected revenue is 20 because the agent will buy both item, for his value for each is at least 10. If each item is offerd for sale at price 22, then the expected revenue is $22 \times (1/2) + 22 \times (1/2) = 22$, which is a bit more. However, if the client offers a "bundle" with both items for a price 32, then the client will get the package with probability 3/4 (when his valuation is not 10 for each item), so that the expected revenue is $(3/4)32 = 24$, which is the maximum possible. Thus, in this example, the "bundle" corresponds to a higher expected revenue.

If the valuations in the previous example are 10 and 50 instead of 10 and 22, then the maximum revenue is to price each item at 50, which corresponds to the expected revenue $(1/2)50 + (1/2)50 = 50$ whereas the bundle priced at 60 gives an expected revenue equal to $(3/4)60 = 45$. Thus, in this case, the maximum expected revenue corresponds to separate prices for the items.

As another possibility, assume that (X_1, X_2) is equally likely to take any of the three values $\{(1, 0), (0, 2), (3, 3)\}$. In this case, the maximum expected revenue corresponds to offering a lottery where the first item is awarded with probability 1/2 for a price 1/2, the second item is offered at price 2, and the pair is offered at price 5. In this case, the agent chooses the option that maximizes his expected surplus

$$\max\left\{\frac{1}{2}X_1 - \frac{1}{2}, X_2 - 2, X_1 + X_2 - 5\right\}$$

and the expected revenue is 2.5.

As these examples suggest, finding the optimal combination of prices and lotteries is hard, even for a single buyer.

These examples are from [Hart and Reny 2012].

More General Single Item Auctions

Relaxing the assumptions of Theorem 3.1 leads to complicated optimal auctions. For example, in network applications we often assume that the utility functions are not linear, but concave, i.e., risk averse. In an ascending auction such as the English auction, the seller's revenue is unchanged since it is still optimal for each bidder to drop out when its private valuation of object has been reached.

Thus seller revenue is unchanged. However, in a first price sealed bid auction the bidders optimal strategy changes. Intuitively, risk aversion leads bidders to be more sensitive to losing and therefore to submit larger bids relative to the risk neutral case.

To understand this somewhat surprising behavior for the first price sealed auction, consider a simple model in which there are two bidders and the reservation price of bidder i is either v_L or v_H where $v_H > v_L > 0$. Further, suppose that both bidders are of the same type, and that p_H and $p_L = 1 - p_H$ denote the probabilities that they bid v_H and v_L, respectively. If both buyers have identical final bids, the winner is decided by tossing a fair coin. This model is from [Maskin and Riley 1985].

In a first price sealed auction, a v_L bidder will bid exactly v_L but what will a v_H bidder do? While we will not prove this, it seems intuitive that the best thing to do is to follow a mixed strategy, i.e., to randomly choose a bid based on a probability distribution. Let $F(b)$ be the c.d.f of the optimal bid when the utility, U, of each buyer is equal to the surplus and let F_R be the c.d.f. when the U is strictly concave. Clearly, $F(v_H) = F_R(v_H) = 1$.

At equilibrium, any bid b the v_H bidder makes must generate the same payoff. When the bidders are risk neutral:

$$(v_H - b)\,(p_L + p_H F(b)) = (v_H - v_L)p_L. \tag{3.4}$$

When the bidders are risk averse:

$$U(v_H - b)\,(p_L + p_H F_R(b)) = U(v_H - v_L)p_L. \tag{3.5}$$

From (3.5):

$$\frac{U(v_H - v_L)}{U(v_H - b)} = \frac{p_L + v_H F_R(b)}{p_L}.$$

Since U is strictly concave:

$$\frac{U(v_H - v_L)}{v_H - v_L} < \frac{U(v_H - b)}{v_H - b}$$

$$\frac{U(v_H - v_L)}{U(v_H - b)} < \frac{v_H - v_L}{v_H - b}.$$

Thus, from (3.4) and (3.5):

$$\frac{p_L + p_H F_R(b)}{p_L} < \frac{p_L + p_H F(b)}{p_L},$$

i.e.,

$$F_R(b) < F(b), \quad b \in (v_L, v_H).$$

This establishes that the optimal strategy is to bid more when the buyers are risk averse, and by implication the expected revenue to the seller is strictly higher in a sealed first price auction than in an English auction.

It turns out that for risk averse buyers even the first price sealed bid auction is not optimal. Optimal auctions generally involve increasing the risk to losing bids further by the imposing payments, and decreasing the risk to high bids by subsidizing them. Thus, the optimal auction can be very complicated and unlikely to be implemented in practice [Maskin and Riley 1984].

When risk aversion is mild, a way to increase the risk for losing bids is to charge a fee for each bid. It can be shown that in a first price sealed bid, the expected revenue to the seller is greater when the bidders do not know how many other bidders there are [McAfee and McMillan 1987] and so under some assumptions it is in the sellers interest to conceal the total number of bidders.

Finally, we address the issue of dependent f_i's. In this case, the form of the optimal auction is odd since the seller must make a net payment to one or more of the bidders. The following example from [Myerson 1981] illustrates this.

Example 3.4 (Dependent Valuations) Suppose $N = 2$ and each bidder values the object with equal probability at 100 or 10. However, the joint distribution is as follows:

$$Pr(10, 10) = Pr(100, 100) = \frac{1}{3},$$

$$Pr(10, 100) = Pr(100, 10) = \frac{1}{6}.$$

Assume that $v_0 = 0$. First consider a Vickrey Auction which is strategyproof even in this setting. The seller's expected revenue is 40. Another feasible auction is to award the item to the highest bidder (breaking ties randomly) at 100 if the bid is 100 and to keep the object unsold otherwise. Then the expected revenue to the seller is 66.67.

The optimal auction actually yields the seller a revenue of 70 and takes the following form. Award the object to the highest bidder (breaking ties randomly). If both bids are 100 then the winner pays 100 and the loser pays nothing. If one is 100 and the other 10, the winner pays 100 and the loser must pay 30. If both are 10 then SELLER pays the winner 5 and the loser 15. The reason the seller must pay this money is to ensure that the expected surplus to each buyer is non negative. In this case, it is exactly zero. Note that if $v_1 = 100$, but agent 1 lies and reports 10 instead, his expected surplus is -30 with probability 2/3 (when $v_2 = 100$). With probability 1/6, $v_2 = 10$ and agent 1 wins giving him a surplus of 95, and with probability 1/6, $v_2 = 10$ and agent one loses, giving him a surplus of 15. Thus, when $v_1 = 100$, agent 1's expected surplus if he lies is negative. Since he has no incentive to lie if $v_1 = 10$ and the bidders are symmetric, the auction is feasible.

Single item auction theory is a highly developed and interesting field and we have barely scratched the surface here. Excellent surveys are found in [McAfee and McMillan 1987] and [Klemperer 1999].

3.4 SPECTRUM AUCTIONS

A nation's spectrum is usually treated as a public good, and is regulated by various government agencies. It is almost always allocated by frequency. Figure 3.2 shows the current allocation of the 3 KHz to 3 GHz frequency band to various "radio services" in the United States. Within a given service such as "TV Broadcasting," various service providers bid for exclusive use within specific geographic areas.

A natural way to assign spectrum rights is by awarding licenses by auction. Each license corresponds to a particular frequency band and geographic region. Thus, a wireless operator is

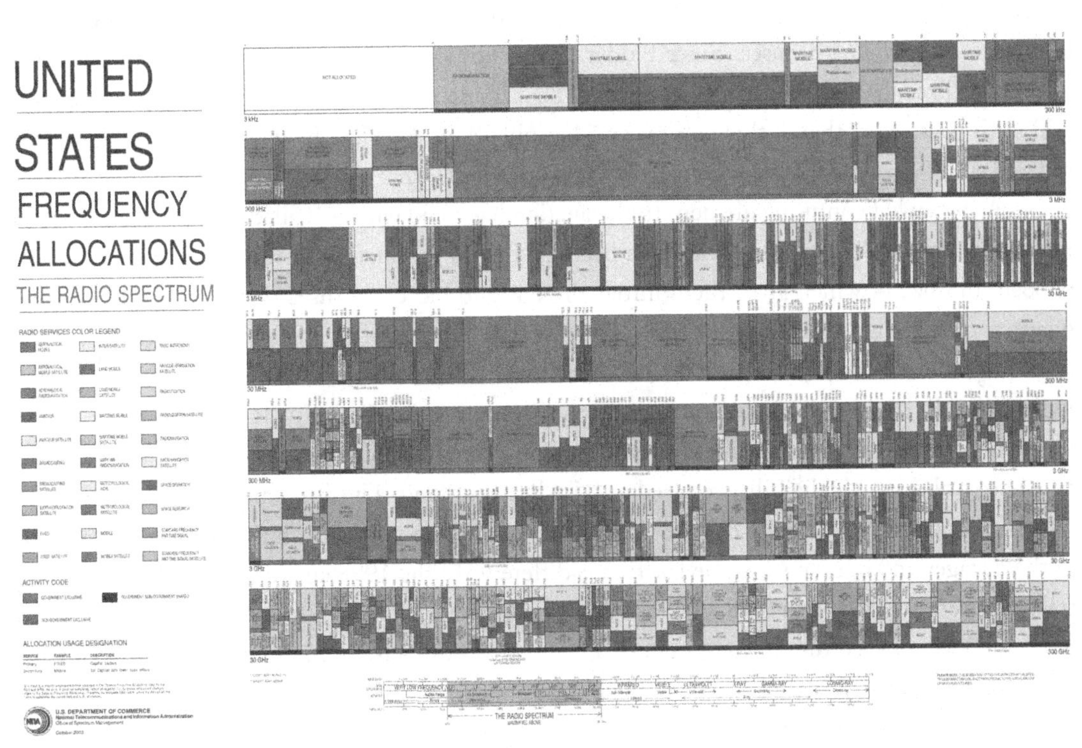

FIGURE 3.2: Spectrum Allocation in the United States.

more likely to be interested in packages of licenses, and the auction is inherently combinatorial. Unfortunately, computational barriers have forced auction designers to seek other ways.

While in our treatment so far we have focused on maximizing the revenue to the seller, the primary goal of a spectrum auction is to ensure that the utility of the buyers is jointly maximized.

The U.S. and other countries have settled on a Simultaneous Ascending Multiple Round Auction (SMRA). Think of an ascending English Auction in which the price is raised in rounds. The auction terminates when there are no new bids **on any one** of the licenses. Thus, the auction starts and ends simultaneously for all the licenses. When the auction ends, each license is sold to its highest bidder, at the highest bid price.

At the beginning of each round, the current high bid for each block is posted (along with the identity of the highest bidder). Next round minimum bids for each item are also posted, so that the auction terminates in a reasonable number of rounds.

Bids are not binding in that they can be withdrawn; however, a penalty is imposed for each withdrawn bid. The penalty is equal to the difference between the value of the withdrawn bid and the value of the winning bid (if this number is negative, no penalty is levied). To see why such a rule makes sense suppose there are two licenses A and B, and two bidders 1 and 2. Also, $U_1(\{A, B\}) = 100$, $U_1(\{A\}) = U_2(\{B\}) = 0$, $U_2(\{A\}) = 110$, $U_1(\{B\}) = 10$, $U_2(\{A, B\} = 175$. Recall that bids can only be made on individual licenses. Since bidder 1 does not know U_2 he will continue bidding for B until he realizes that he although he can win B he is going to lose A since user 2 has bid more than 100 for it. Let us say that at this point user 1 has bid 30 for B. Then he will want to withdraw his bid and suffer a penalty. If the winning bid ends up being 5, this penalty is 25.

A somewhat confusing element to the auction has to do with so-called Activity Rules, which incentivize the bidders to bid for multiple licenses. Before the auction starts, each bidder puts up a deposit, the size of which gives them initial eligibility to bid on licenses that total up to a certain quantity of spectrum (measured in Mz). The larger the deposit the greater the eligibility. At the beginning of each round, a bidder is allowed to make a bid for any license for which it holds the highest bid. In addition it can bid on licenses that total up to its eligibility. If the total quantity c of spectrum bid on in any round is **less** than a fraction f of the bidder's eligibility, then the bidder is penalized by having his eligibility reduced to $\frac{c}{f}$ in the next round. The fractions f are predetermined, but the auctioneer can actually change them twice during the auction. Thus, the auction proceeds in three stages and in each stage the fraction is different. In the 1998 auction in 220 Mz range, the fractions were 0.8, 0.9, and 0.98. As an example, suppose a bidder has eligibility of 100 Mz, and bids on 40 Mz of spectrum in some round. Then if $f = 0.8$ then his eligibility will drop to 50 Mz in the next round. If $f = 0.98$ then eligibility drops to about 40.8 Mz.

As pointed out in [Milgrom 1999], the activity rules are important to ensure that bidders do not delay their bidding. Assume that each bidder is budget constrained, i.e., there is an upper limit to the amount of money he can spend on the auction. Then, it is in a bidder's interest to delay his bids on a given license as he waits for his competitors to exhaust their budgets on licenses that he does not want. Thus, the auction is likely to proceed very slowly without a mechanism similar to the activity rules.

Simultaneous Multiple Round Auctions are not perfect. While they provide a wealth of information to the bidders, they also open the door to alliances and collusion. While collusion has been detected in some auctions, the effect on prices has been deemed to be negligible. See [Cramton and Schwartz 2002] for more details.

3.5 APPENDIX: PROOF OF MYERSON'S THEOREM

Theorem 3.1 is a classical result. It is proved in [Myerson 1981] and in most economics books. We repeat a proof here, based on Myerson's paper, hoping to add some clarity.

Recall that the users i have independent valuations $U_i \in [a_i, b_i]$ with a positive pdf $f_i(x)$ on $[a_i, b_i]$ and cdf $F_i(x)$ such that the virtual valuation

$$c_i(x) := x - \frac{1 - F_i(x)}{f_i(x)}$$

is non-decreasing on $[a_i, b_i]$.

THE FOUR STEPS

The proof uses the following steps: (1) a principle; (2) a characterization of incentive compatible auctions; (3) a calculation; and (4) a verification. We first outline the steps and then provide details.

Step 1

The principle is the *revelation principle* that says that when we design an auction, we can assume that the agents reveal their true valuation to the auctioneer. Indeed, say that the best strategy for the agent is to bid $\phi_i(U_i)$ when his valuation is U_i. In that case, one can assume that the agent reveals U_i to the auctioneer who then computes $\phi_i(U_i)$. In other words, we design the optimal auction by computing the best function $\phi_i(U_i)$ instead of the best function $\psi_i(b_i)$ of a bid $b_i = b_i(U_i)$ that the agent computed; this is more general.

Step 2

The characterization of incentive compatible auctions is stated in the following theorem. Say that user i gets the item with probability $\pi_i(\mathbf{U})$ and pays $p_i(\mathbf{U})$ when the valuations are $\mathbf{U} = (U_1, \ldots, U_N)$. Let $p_i(u) = E[p_i(\mathbf{U})|U_i = u]$ and $\pi_i(u) = E[\pi_i(\mathbf{U})|U_i = u]$.

Theorem 3.2 An auction is incentive compatible if and only if $\pi_i(u)$ is non-decreasing in u and

$$p_i(u) = u\pi_i(u) - \int_{a_i}^{u} \pi_i(x)dx - S_i(a_i) \tag{3.6}$$

for some constant $S_i(a_i)$ that is then the expected net surplus of user i when his valuation is a_i.

Note, in particular, that the price is essentially determined by the probability of winning. For instance, if the item goes to the agent with the largest valuation and if agent i does not get any surplus when his valuation is a_i, then $p_i(U)$ is determined. It follows that all auctions with these characteristics achieve the same revenue. This fact is called the *revenue equivalence theorem.*

Step 3

The calculation shows, using (3.6), that the revenue of the auction is

$$E\left(\sum_i p_i(\mathbf{U})\right) = E\left[\sum_i \pi_i(\mathbf{U})c_i(U_i)\right] - \sum_i S_i(a_i). \tag{3.7}$$

Step 4

The verification consists in showing that the auction described in Theorem 3.1 is individually rational and maximizes (3.7) while satisfying the characterization of Theorem 3.2 of incentive compatible auctions.

Proof of Theorem 3.2.

1. Necessity. Assume that the auction is incentive compatible. The expected surplus of agent i bidding x while his valuation is U_i is given by

$$\begin{aligned} U_i\pi_i(x) - p_i(x) &= x\pi_i(x) - p_i(x) + (U_I - x)\pi_i(x) \\ &= S_i(x) + (U_i - x)\pi_i(x), \end{aligned} \tag{3.8}$$

where $S_i(x) = x\pi_i(x) - p_i(x)$ is his expected surplus when his valuation is x and he bids his valuation. We want to design an auction where the users have an incentive to bid their true valuation. That is, we want the expression (3.8) to be maximized by $x = U_i$. In particular,

$$S_i(U_i) = \max_x\{U_i\pi_i(x) - p_i(x)\},$$

so that $S_i(U_i)$ is the maximum of linear functions of U_i (one linear function for each x) and is therefore a convex function of U_i.

For the expression to be maximized by $x = U_i$, the derivative of (3.8) with respect to x should be equal to zero for $x = U_i$. This implies that

$$S_i'(U_i) = \pi_i(U_i).$$

In particular, $\pi_i(U_i)$ is increasing in U_i since it is the derivative of a convex function. Hence,

$$S_i(U_i) = S_i(a_i) + \int_{a_i}^{U_i} \pi_i(x)\,dx.$$

Finally,

$$p_i(U) = U\pi_i(U) - S_i(U) = U\pi_i(U) - \int_0^U \pi_i(x)dx - S_i(a_i),$$

as claimed.

2. Sufficiency. Assume that the conditions of the theorem are satisfied. We show that the auction is incentive compatible.

   Say that agent i declares a valuation x. His surplus is then

$$U_i\pi_i(x) - p_i(x).$$

The derivative with respect to x is equal to

$$h_i(U_i, x) := U_i\pi_i'(x) - p_i'(x).$$

Using (3.6), we find that

$$p_i'(x) = x\pi_i'(x).$$

Hence,

$$h_i(U_i, x) = U_i\pi_i'(x) - x\pi_i'(x) = (U_i - x)\pi_i'(x).$$

Since $\pi_i(x)$ is non-decreasing, by assumption, we see that $h_i(U_i, x)$ is nonnegative for $x \leq U_i$ and non-positive for $x \geq U_i$. We conclude that $U_i\pi_i(x) - p_i(x)$ is maximized for $x = U_i$, so that the auction is incentive compatible. ∎

CALCULATION

We show (3.7). Using (3.6), we find that the revenue R of the auction is

$$R := \sum_i E\left[p_i(\mathbf{U})U_i - \int_{a_i}^{U_i} \pi_i(u)du\right] - \sum_i S_i(a_i).$$

Now,

$$E\left[\int_{a_i}^{U_i} \pi_i(u)du\right] = \int_{a_i}^{b_i} f_i(v) \int_{a_i}^{v} \pi_i(u)dudv = \int_{a_i}^{b_i} \pi_i(u)(1 - F_i(u))du$$

$$= \int_{a_i}^{b_i} \pi_i(u)\frac{1 - F_i(u)}{f_i(u)} f_i(u)du = E\left[\pi_i(U_i)\frac{1 - F_i(U_i)}{f_i(U_i)}\right].$$

(The second equality follows by changing the order of integration.)

Hence,

$$R = \sum_i E\left[p_i(\mathbf{U})(U_i - \frac{1 - F_i(U_i)}{f_i(U_i)}\right] - \sum_i S_i(a_i)$$

$$= \sum_i E[p_i(\mathbf{U})c_i(U_i)] - \sum_i S_i(a_i),$$

as claimed.

VERIFICATION

The Myerson auction is such that $\pi_i(\mathbf{U})$ is one if $c_i(U_i) \geq c_j(U_j), \forall j \neq i$ and $c_i(U_i) \geq 0$ and in that case, $p_i(\mathbf{U})$ is the smallest value of x such that $c_i(x)$ has those properties. We want to show that the auction is individually rational, incentive compatible, and maximizes the auctioneer's revenue.

The auction is individually rational because the payment $p_i(\mathbf{U})$ is always less than or equal to U_i, by construction, so that the user surplus is nonnegative.

To show that the auction is incentive compatible, we first note that the auction is such that $\pi_i(\mathbf{U})$ is nondecreasing in U_i since $c_i(U_i)$ is assumed to be nondecreasing and the event that $c_i(U_i)$ exceeds 0 and $c_j(U_j)$ for $j \neq i$ increases in U_i. It remains to show that (3.6) holds.

Observe that

$$\pi_i(\mathbf{U}) = 1\{U_i \geq p_i(\mathbf{U})\}.$$

Hence,

$$U_i\pi_i(\mathbf{U}) - \int_{a_i}^{U_i} \pi_i(x, \mathbf{U}_{-i})dx = U_i \times 1 - \int_{a_i}^{U_i} 1\{x \geq p_i(\mathbf{U})\}dx$$

$$= U_i - (U_i - p_i(\mathbf{U})) = p_i(\mathbf{U}).$$

Thus, (3.6) holds with $S_i(a_i) = 0$. Note that the minimum value of x such that $c_i(x)$ is positive is at least a_i, so that the minimum value of p_i is a_i, so that $S_i(a_i) = 0$.

Finally, the Myerson auction maximizes the expression (3.7) because it maximizes the term in the expectation and it is also such that $S_i(a_i) = 0$ for all i and any individual rational auction is such that $S_i(a_i) \geq 0$.

3.6 SUMMARY AND REFERENCES

Auction Theory is a standard topic in Economics. We decided to discuss auctions in this text because they are one of the main mechanisms for allocating resources.

The observation that a second price auction is incentive-compatible is due to Vickrey [Vickrey 1961]. This auction was generalized to the VCG mechanism in [Clarke 1971] and [Groves 1975]. In a VCG auction, the users reveal their true utility and the allocation maximizes the sum of the utilities. Unfortunately, implementing such an auction is both informationally and computationally complex.

Designing a revenue-maximizing auction is still an open problem for most situations. Myerson [Myerson 1981] solved the problem for a single item. We reviews that elegant result in the "regular" case where the virtual valuations are monotone in the true valuations. The general case of multiple items is still open and is known to be computationally hard; see [Krishna 2009] for a presentation of Auction Theory.

Another interesting application of auction theory is the placement of advertisements in search engines such as Google. For example, see [Edelman et al. 2005].

CHAPTER 4
Matching

In a network, different users or applications require sets of resources, such as bandwidth or servers. Often the network has to select a "match" between users or applications, that we call *agents*, and resources. Resource allocation problems can often be cast in terms of matching pairs of entities. Matching Theory itself is a rich branch of discrete mathematics [Plummer and Lovász 1986]. Here we will focus on two special cases.

1. The agents have declared preferences on the resources they would like to be matched with. The goal of the matching is to accommodate these preferences as well as possible, in a sense to be made precise. For instance, some tasks can be performed by different servers with varying performance metrics.
2. The agents have declared utility functions of some variable whose value is determined by the matching. For example, suppose a router has to match the packets at its input ports to output ports. Then the flows to which these packets belong could declare utility functions in terms of the rate their packets receive under the router's matching algorithm. The goal of the matching algorithm might be to maximize the sum of the flow utilities.

To make these ideas more clear, consider the following examples, each of which we expand upon in subsequent sections.

HOUSING

There are N people and N houses. Each person has a strict ranked preference order for the N houses and initially owns a house. How can we trade the houses among the people? This is an example of a one-sided matching problem since people have preferences but houses do not. Clearly, it may not be possible to assign each person her preferred house. However, a desirable matching should have two basic properties that we define next.

Definition 4.1 (Stable Matching) A matching is *stable* if no two persons would be better off by trading among themselves.

Thus, if a matching is such that Alice and Bob would rather exchange their houses, the matching is not stable. Clearly, a matching that is not stable is not desirable.

Definition 4.2 (Core) A matching is in the *core* if no subset of the people would all be better off by trading among themselves instead of accepting the matching.

Thus, if a matching is such that a group of agents, say Alice, Bob, and Charles, would have all been better off by trading among themselves instead of with the allocation assigned by the matching, then this matching is not in the core. Only matchings that are in the core are desirable.

The TTCA algorithm that we explain below produces a stable matching that is in the core.

MARRIAGE

A familiar matching problem is to pair men and women. There are N men and N women. Each man ranks the women in order of preference and each woman ranks the men in order of preference. The goal of the matching is to accommodate these preferences. This is an example of a two-sided matching problem. Again, the goal is to find a stable matching, i.e., a matching such if any two sets of matched partners were to swap partners at least one of the four people would be worse off.

SCHOOL ADMISSIONS

N students apply to M graduate schools. For $m = 1, \ldots, M$, school m can admit S_m students. Each student has a ranked preference of the M schools and each school has a ranked preference of the students. Which students should each school admit? This is an example of a many-to-one matching problem since many students may be matched to a given school. The notion of stability is similar as before.

SWITCHING

Flows arriving at input port i of a router must be switched to output port j to ensure that the packets get to their destinations with a maximum throughput. The router is fast enough to deliver, in one time slot, at most one packet to each of the output ports. Then the router is solving a matching problem at each time slot. The preference functions of the input and output ports are not obvious, but as we will see, by defining them cleverly we can create a router with optimal throughput.

4.1 HOUSING MATCHING

In this section we show how to find a stable matching that is in the core and *strategyproof*. A matching algorithm is strategyproof if no one can end up with a better house by lying about his/her ranked order. We first clarify the notions of stability and core.

STABILITY AND CORE

Figure 4.1 illustrates the case of three persons A, B, C and three houses 1, 2, 3. Initially, A owns house 1, B owns 2, and C owns 3. In the left part of the figure, arrows indicate the preferences. For

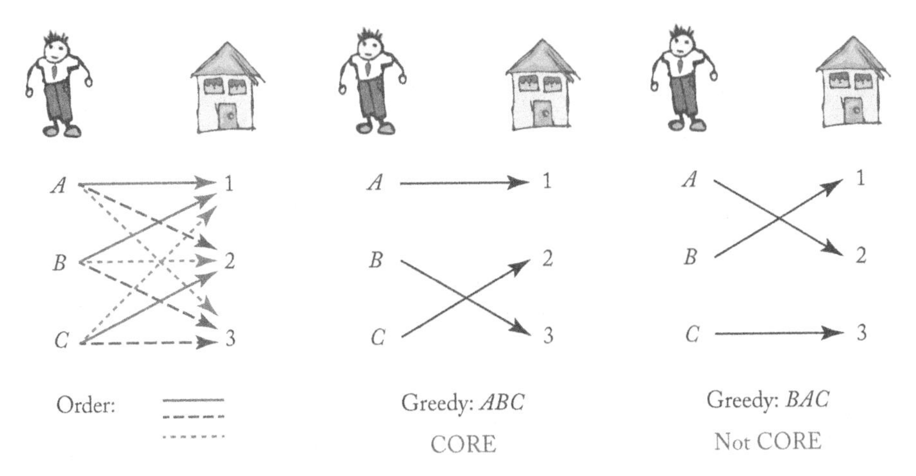

FIGURE 4.1: Examples of matchings.

instance, A has a strict decreasing order of preferences $(1, 2, 3)$. The order for B is $(1, 3, 2)$ and that for C is $(2, 3, 1)$. The middle part of the figure shows the matching that results from A choosing the house he prefers, then B selecting the house he prefers after A has made his choice, and finally C picking the remaining house. This matching corresponds to a greedy algorithm when the persons act in the order ABC.

Such a greedy matching is always stable. Indeed, A does not want to swap with anyone since he got his preferred house. Also, B does not want to swap with C. Thus, any greedy matching, for any given order of selections, is stable. The right-hand part of the figure shows the greedy matching when the persons select in the order BAC.

To understand the concept of core, consider the matching on right of the figure. Although this matching is stable, it is not acceptable to A. Indeed, A would prefer not to take part in the exchange and keep his original house 1 that he prefers to house 2 he gets in this matching. Thus, in that matching, there is a subset of the persons, namely $\{A\}$, that prefers not to participate in the matching.

More generally, a matching is in the core if there is no subset S of the agents such that all the agents in the subset would be better off by trading among themselves than they are in the matching. The matching in the middle part of Figure 4.1 is in the core. To see this, note that each agent gets a house that is at least as preferable as his original house. Also, $\{A, B\}$ would not be better off trading among themselves, because A could get 1 and B would get 2, which he finds less preferable than house 3 he gets in the matching. The same is true for the other subsets.

Thus, a matching that is not in the core is not acceptable and would not happen in practice. Summing up, it is easy to find stable matchings. However, in a collaborative mechanism such as the

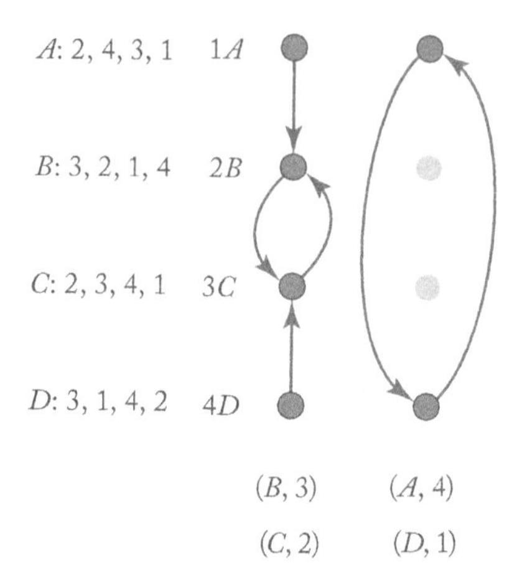

FIGURE 4.2: An example of the TTCA.

house matching, an allocation is acceptable only if it is in the core. We revisit this concept in the next chapter.

TOP TRADING CYCLE ALGORITHM (TTCA)

The *top trading cycle algorithm* (TTCA), generally attributed to Gale, produces a matching that is stable, strategyproof, and in the core. We first illustrate the algorithm in the case of four people (A, B, C, D) and four houses $(1, 2, 3, 4)$. The left-hand part of Figure 4.2 shows the preference orders of the four people for the four houses. Thus, Alice (A) ranks the houses in decreasing order $2, 4, 3, 1$, and similarly for the other people. The left-most graph has four nodes. Each node mX represents both one person X and one house m, as indicated. An arrow from node mX to node nY means that person X's top preference is currently house n. For instance, the arrow from $1A$ to $2B$ indicates that the top preference of A is 2. Similarly, the arrow from $2B$ to $3C$ indicates that the top preference of B is 3, and so on. This graph has a cycle: $2B \to 3C \to 2B$. TTCA assigns each person in the cycle his/her top choice. Thus, B gets 3 and C gets 2, as indicated below that graph. One then removes the two people (B, C) and the houses (2, 3) that have been matched. The next graph shows the updated situation with only two nodes $1A$ and $4D$ and the arrows corresponds to the top preferences of A and D after 2 and 3 are removed. This graph again has a cycle and the persons A, D in that cycle are assigned their top preference, as shown under the graph. Thus, TTCA produces the matching $\{(A, 4), (B, 3), (C, 2), (D, 1)\}$.

Figure 4.3 shows the TTCA algorithm for a different set of preferences.

Summing up, in the case of N people and N houses, TTCA starts with a graph that has N nodes. Each node, say mX, corresponds to a pair of one person X and one house m. The algorithm

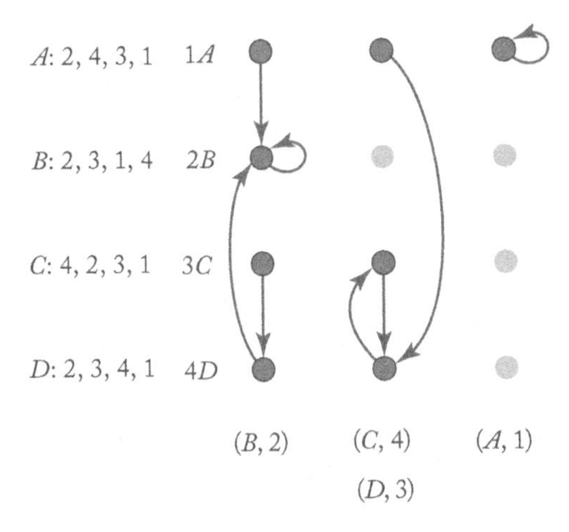

FIGURE 4.3: Another example of the TTCA.

then draws an edge from mX to nY if the top preference of person X is house n. The algorithm then matches the people in a cycle, including self-loops, with their top preference, then removes the matched persons and houses, and continues.

Theorem 4.1 If each agent has strict preferences, there is a unique allocation in the core. TTCA produces that allocation and it is stable and strategyproof.

Proof. First, we show that TTCA produces a matching. Note that each stage produces one or more node-disjoint cycles. It produces cycles (possibly, self-loops) because it has M nodes and M edges. If there were no cycle, then each node would belong to a loop-free path. However, a loop-free path has one edge fewer than nodes. Thus, it is not possible that no cycle exists. The cycles must be node disjoint because each node has only one outgoing edge, so that no node can belong to more than one cycle. Thus, the algorithm makes progress as long as not all persons are matched and the algorithm must then terminate with a complete matching.

Second, we claim that TTCA produces a stable matching. To see this, assume as a contradiction hypothesis that the matching contains $(A, 1)$ and $(B, 2)$ where A prefers 2 to 1 and B prefers 1 to 2. Say that A was assigned house 1 before B was assigned house 2. When A was assigned house 1, that house was A's preferred house among the remaining unassigned houses. Thus, it cannot be that A prefers 2 to 1. Similarly, if A and B are assigned houses at the same time, they must get their preferred house among the remaining ones.

Third, we show that TTCA produces the same matching as any allocation in the core. To see this, first observe that the allocations that TTCA makes in the first step must be the agent's top choices, otherwise those agents would form a coalition and trade among themselves. Also, this allocation must be the same as in any core allocation. In the example shown in Figure 4.2, agents B and C must get their top allocation and any location in the core must be the same, otherwise $\{B, C\}$ would leave the group and trade among themselves. That is, they would block any other allocation. Similarly, the agents that get allocated the second step must get their top choices among the remaining houses, for otherwise they would leave the coalition and block the allocation. Any allocation in the core must also allocate them the same houses. The same argument holds for the subsequent steps.

Fourth, we claim that the TTCA allocation is in the core. Indeed, if it were not, it would be blocked by a subset of the agents. However, these agents are not those who are allocated in the first step, since they get their top choice. Also, these agents are not those that are allocated in the second step. Continuing in this way shows that no subset of the agents blocks this allocation, so that it is in the core.

Fifth, we show that TTCA is strategyproof. The agents that are allocated in the first step cannot gain by lying about their preferences, since they get their top choice. Similarly, the agents that are allocated in the second step cannot gin since they get their top remaining choice. Continuing in this way shows that no agent can gain by lying. ■

4.2 STABLE MARRIAGE

Consider the marriage problem with N men and N women, each having a strict ranked order of the persons of the other sex. Finding a good matching is called a *two-sided matching problem.*

We say that a matching is *stable* if there is no Alice and Bob who would prefer to be matched instead of staying with their current partner. Also, a matching algorithm is *strategyproof* if no one can end up with a better partner by lying about his or her preferences.

Example 4.1 (Multiple Stable Matchings) Consider the following example for $N = 3$. The men are m_1, m_2, m_3 and the women are w_1, w_2, w_3. Their preferences are as follows:

$$
\begin{array}{ll}
m_1\colon\ w_2 > w_1 > w_3 & \qquad w_1\colon\ m_1 > m_3 > m_2 \\
m_2\colon\ w_1 > w_2 > w_3 & \qquad w_2\colon\ m_3 > m_1 > m_2 \\
m_3\colon\ w_1 > w_2 > w_3 & \qquad w_3\colon\ m_1 > m_2 > m_3
\end{array}
$$

There are exactly two stable matchings: $M = \{(m_1, w_2), (m_2, w_3), (m_3, w_1)\}$ and $M' = \{(m_1, w_1), (m_2, w_3), (m_3, w_2)\}$.

GALE-SHAPLEY ALGORITHM (GSA)

Theorem 4.2 shows that the *Gale-Shapley algorithm* give a stable matching. Here is the algorithm.

Gale Shapely Algorithm (GSA):[Gale and Shapley 1962] While there is a man m who has not been assigned, let him propose to the woman who he prefers the most but who has not previously rejected him. If she is currently partnered with a man m' whom she prefers more highly, she rejects the proposal. Otherwise, she severs her matching with m' and partners with m.

In the example above, GSA produces M. Each man is no worse off and each woman no better off under M than under M'. We will see below that this is a characteristic of GSA. If we switch the roles of men and women, i.e., modify the GSA so that women propose to men etc., the resulting matching is M'.

Theorem 4.2

(a) The Gale-Shapley algorithm produces a stable matching.

(b) Among all possible stable matchings, in GSA each man is matched with his preferred partner. In particular, no man has an incentive to lie about his preferences.

(c) Among all possible stable matchings, in GSA each woman is matched with her least preferred partner. In particular, she has an incentive to lie about her preferences and GSA is not strategyproof.

Proof.

(a) First, we show that the algorithm terminates. First observe that once a woman is partnered, she is always in a match and only switches to a preferred man thereafter. Also, each woman must be proposed at least once, since there are as many men as women. Since each woman can only switch finitely many times, the algorithm terminates.

Second, we show that GSA terminates in a stable matching. Suppose it doesn't. Then there must be a pair (m, w) such m prefers w over his current wife w' and w prefers m over her current husband m'. This means that m did not propose to w after she had partnered with m'. This would have to be because m had already partnered with some w'. But when m decided to partner with w' he would have tried to partner with w who obviously was not available because she was partnered with some man m'' who she preferred more than m. But then she could have never broken up with him for m'.

(b) Suppose GSA terminates in the following state: There is a matching (m, w') such that m prefers another woman w more than w'. This means that he had proposed to w but was rejected because w received (from her perspective) a better proposal from some other man, m' who she prefers over m. Thus, w' must be the highest preferred woman of m's preference ordering, such that of the men who prefer her more than m does, she prefers m the most.[1]

1. Let us say that m prefers w more than m' if w is higher in m's preference ordering than in m''s preference ordering.

Thus, there is no stable matching in which m could be matched with a woman higher than w' on his preference list, and he has no incentive to lie about his preference list.

(c) Let $\mathcal{M} = \{\ldots, (m, w), \ldots\}$ be the male-optimal matching that GSA produces. Suppose for the sake of contradiction that there exists a stable matching $\mathcal{S} = \{\ldots, (m^*, w), \ldots, (m, w'), \ldots\}$ such that m^* is lower on w's list than m. We will argue that $\mathcal{S}$ cannot possibly be stable by showing that (m, w) would rather be matched together than stay with their partner in $\mathcal{S}$. By assumption, w prefers m to m^* since m^* is lower on her list. And m prefers w to his partner w' in $\mathcal{S}$ because w is his partner in the male-optimal matching $\mathcal{M}$. This is a contradiction. ■

ON STRATEGYPROOF MATCHINGS

We saw that GSA is not strategyproof. In fact, we have the following result.

Lemma 4.1 No algorithm for the stable marriage problem is stragtegyproof.[Roth 1982]

Proof. Consider again the example we discussed above.

Suppose there is an algorithm that produces the matching M. Suppose that w_1 lies and declares her preferences to be $m_1 > m_2 > m_3$. Then there is exactly one stable matching, M', and w_1 is better off under M' than under M. If, on the other hand, an algorithm produces M' then m_1 can improve his match by stating his preferences to be: $w_2 > w_3 > w_1$ which will result in only one stable matching: M. Since there is no strategyproof algorithm for the example, we have proven the result in general. ■

Since GSA is not strategyproof, it is natural to ask what can happen if one someone lies. It turns out that if the resulting matching is a Nash Equilibrium in which no participant can strictly improve their match by lying, then the resulting matching is still stable relative to the true preferences!

The following lemma discusses *weakly dominated* strategies. In a game, the strategy of one player is weakly dominated if there is another strategy that yields rewards that are as large, for every possible strategies of the other players.

Lemma 4.2 Every Nash Equilibrium at which no weakly dominated strategy is employed yields a stable matching relative to the true preferences [Roth 1984].

Proof. Suppose GSA yields a matching M based on stated preferences, and that these stated preferences form a Nash Equlibrium with no weakly dominated strategies (no further lying by a participant can yield a strictly better match for them), and contrary to the theorem, M is unstable relative to the true preferences. Then there is a pair (m, w) not in M, where m prefers w to his partner and w prefers m to her partner. Now since m has no incentive to lie, he must have proposed to w under GSA but was rejected by her. But now let us modify the stated preferences of w where she ranks m higher than all the other men, but leaves her other preferences the same (all the other men and women do not change their stated preferences). If GSA is run on this instance of the problem, w will accept m when he proposes and (m, w) will be a match. Thus, by unilaterally changing her preferences w has strictly improved her match and therefore the preferences that yield M could not be a Nash Equilibrium. This contradicts our assumption. ■

As we will see in the next section, in the context of applications it is frequently the case that the preference relations are not complete, there are some men (women) that a woman (man) would never want to marry. Thus, a subset of the pairs (edges of the bipartite graph) are eligible to be in a matching. In this case it is easy to extend our notion of stable matchings: if (m, w) and (m', w') are in such a matching then they both are eligible pairs and if m prefers w' over w, then w' must prefer m' over m, etc.

First, observe that if we run GSA on such a problem, it will find a stable matching but the matching may not be *perfect*, i.e., there might be some pairs of nodes that remain unmatched.

The natural question to ask is the following: *Does an instance of a bipartite graph with incomplete preferences contain a stable perfect matching?* To answer, we construct an *associated system* by adding a fictitious man m, called the widower, and a fictitious woman, w, called the widow. Now, induce a complete set of preferences by retaining the original (incomplete) preferences and adding to them as follows: each man (woman) lists the widow (widower) after their genuine (original) preferences and then lists, in some arbitrary order, the women (men) that he (she) will never marry. Finally, the widow (widower) can have arbitrary preferences as long as she (he) lists the widower (widow) last.

Theorem 4.3 The original incomplete problem has a stable perfect matching iff the associated problem has a stable matching in which the widow (w) and widower (m) marry.

Proof. Now suppose that GSA terminates with a matching that includes the pair (m, w), and there is a man who is paired with a woman who he was not willing to marry in the original system. By construction he prefers w to this woman, and w prefers him to m. So the matching cannot be stable. Thus, if the associated problem has a stable matching that includes (m, w) as a pair, there is a stable matching in the original problem. Now suppose that GSA terminates and (m, w) is not a pair. Observe that any stable matching in the original problem can be converted to one in the associated problem by adding the matching (m, w). Thus, there cannot be stable matching in the original problem if GSA terminates without matching the widow and widower. ■

We can strengthen the theorem somewhat by observing that if we apply GSA to the associated problem (the men propose), we know that each woman will get their least preferred eligible partner.

Thus, a stable perfect matching exists in the original problem iff GSA yields a stable perfect matching in the associated problem in which (m, w) are paired.

4.3 SWITCHES AND MATCHINGS

Figure 4.4 represents an N by N switch. This switch transfers fixed-size packets called *cells*. For convenience divide the operation of the switch into the following $s + 2$ sequential phases that collectively take one time slot to complete.

1. Arrival: At most one fixed size cell arrives at each of the input ports. Let $E(c)$ be the destination output port for cell c.
2. s different scheduling phases: In each Scheduling phase, at most one cell is transferred from each input port to the corresponding output port. No output port can receive more than 1 cell per scheduling phase. Thus, the scheduling phases can be thought of as s matchings between input and output ports. We refer to s as the speedup factor of the switch.
3. Departure: For each output port with a non empty buffer, exactly one cell from its buffer leaves the switch.

When $s = N$, the input buffers are always empty and all of the buffering is done at the output. This allows one to think of the switch as comprising of N independent work conserving multiplexers. These so-called **Output Queued** switches can implement service disciplines such as First-Come-First Served and Weighted Fair Queuing to control performance variables such as latency and loss. However, their high-speed up factor makes them expensive to build.

When $s = 1$ all the queueing is at the inputs and the switch is called **Input Queued**. These switches may be able to provide 100% throughput but may delay cells for a long time in the input buffers.

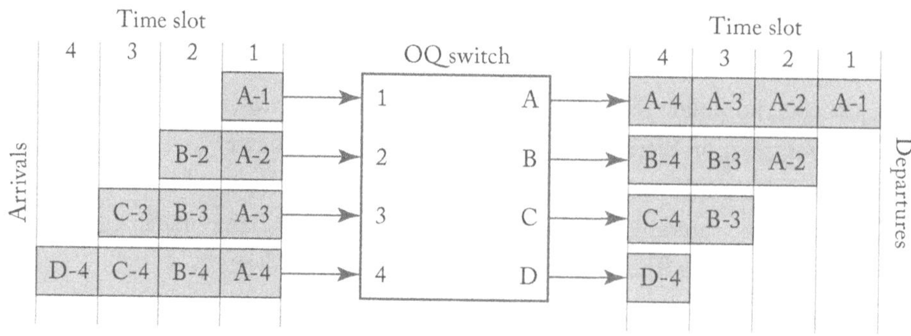

FIGURE 4.4: Example: Traffic Pattern that requires $s = 2 - \frac{1}{N}$. The figure also shows the departures for a FCFS output queued switch.

When $1 < s < N$ the switch is **Input-Output Queued** and there are buffers at both the input and output port buffers. An interesting question one can ask is the following: What is the minimum value of s for which a switch can perfectly emulate an output queued switch? For example, for any pattern of cell arrivals and an output queued switch S^o, we would like to design a switch with speed up $s < N$ whose departures are indistinguishable from S^o. We allow S^o to implement any Monotone Scheduling Policy at its output ports. A Monotone scheduling policy is one such that has the following ordering property:[2]

> Let cells c and c' be in the system at time τ, and suppose that c departs the system before c' if no more cells arrive after τ. Then c must depart before c' regardless of any arrival pattern after time τ.

The reason we can hope to emulate S^o is the following. Consider a cell that arrives at the input queue of the switch at time t_0, is switched to the output port at time t_1 and begins being served by the output queue at time t_2. In an output queued switch, $t_1 = t_0$, and so it waits in the output queue for time $q_o = t_2 - t_1$. Now suppose we had an input-output queue that delivered the cell to the output after t_0 but before t_2. Then, it would be possible for the switch to ensure that the cell left by t_2. If we can do this for each cell, then the output switch can be emulated exactly.

Theorem 4.4 It is possible to emulate any output queued switch S^o with an Input-Output Queued switch of speedup 2 as long as S^o implements a monotone scheduling policy. [Chuang et al. 1999]

Proof. When cell c arrives at port i and is destined for port p, it is marked with the time, $F(c)$ that it would depart under S^o if no more cells were to arrive. We refer to F as the finishing time. Note that $F(c)$ can increase or decrease with time. Cell c is also added to two priority lists.

- Input i Priority List: c is placed at a particular (to be specified) position in the list. Once c has been transferred to its output it will be removed from the list.
- Output p Priority List: All the cells on the **input side** of the switch that are destined for p are ordered by their finishing times.

Clearly, there are N input priority lists and N output priority lists.

In each Scheduling phase, preference relations for each of the input and output ports are derived from the input and output priority lists, and then GSA is run to find a stable matching. The preference relations are defined as follows.

- For input i, $p \succ_i q$ iff a cell destined for port p is higher in its priority list than one destined for q.
- For output port p, $i \succ_p j$ iff a cell destined for p from i appears higher in p's priority list than one destined from j.

2. See [Georgiadis et al. 1997] for other interesting properties for these scheduling disciplines

The stable matching defines a schedule, i.e., if input port i is matched to output port p then transfer the highest input priority ordered cell at i which is destined for p.

In the Departure phase, each output port serves the cell in its buffer that has the smallest finishing time.

From the preceding discussion it is clear that if we specify how to insert each arriving cell into its Input Priority List we have completely specified the operation of the switch: For any cell c at input port i and destined for port p, let its OutputWait(c) be the number of cells at p with smaller finishing times than $F(c)$. Also let InputWait(c) be the number of cells in i's input priority list. Finally, based on the intuition above, let Slack(c)=OutputWait(c) - InputWait(c). If Slack(c)>0, there is enough time for the switch to serve c at the same time that it would leave S^o. Our rule for inserting into the Input Priority List is simply this:

> If c arrives at input port i, place it on the OutputWait(c)+1th position on the Input Priority List so that Slack(c)=0. If there are fewer than OutputWait(c) cells at i, place it last in the list so that Slack(c)>0.

Now let us see what happens to Slack(c) for any cell c in input queue i destined for output port p. At the end of each Scheduling phase, since the matching is stable, at least one of the following must be true.

- c is transferred to p. In this case c is no longer on the input and therefore Slack(c) is not defined. It cannot happen that both i is not served and p is not fed in a scheduling phase, because then they can be matched.
- A cell higher in Input Priority List i is transferred: In that case InputWait(c) goes down by 1 and OutputWait(c) goes up by at most 1. Thus, Slack(c) goes up by at least 1.
- A cell higher in Output Priority List p is transferred: Then InputWait(c) is the same and OutputWait(c) goes up by 1. Thus, Slack(c) increases 1.

Thus, if c is still on the input side of the switch at the end of the two Scheduling Phases Slack(c) would have increased by 2. In an arrival phase, InputWait(c) goes up by at most 1. In a departure phase OutputWait(c) must go down by exactly 1. Thus, the net change to Slack(c) in one time slot is non-negative. Since Slack(c) starts out by being non-negative it must remain so until OutputWait(c)=0. When this happens InputWait(c) must be zero as well (by definition of Slackness), implying that either c is already transferred (in which case it will be selected to leave the switch in the Departure phase), or it is the highest ranked cell in its input priority list and its output priority list, and therefore must be matched to its destination by GSA in a Scheduling phase of the slot and will picked in the Departure Phase. ■

We now show that Theorem 4.4 is a strong result by establishing that there is an arrival pattern such that if S^o is an NXN first-come-first-served switch, a speedup of $2 - \frac{1}{N}$ is required. The arrival pattern is shown in Figure 4.4 and is from [Chuang et al. 1999].

Assume that the FCFS output schedular breaks ties by picking the cell that came from the lowest indexed port. We leave it to the reader to show for the arrival pattern in the figure, at least 7 matchings must be done (2 in time slots 1,2,3 and one in time slot 4) to emulate the output queued switch. Now the pattern of arrivals can be repeated indefinitely. We need $s \geq \frac{7}{4} = 2 - \frac{1}{4}$ for $N = 4$. It is straightforward to generalize to arbitrary N.

4.4 MAXIMUM WEIGHT MATCHING

We once again consider the scheduling of an input-output queued switch. We review the instability of the maximal matching algorithm, the stability of the maximum weighted matching and of a randomized algorithm based on queue lengths.

Example 4.2 (Switch) Consider the situation shown in Figure 4.5. This switch has three active inputs for two outputs. The rate of the flow from input i to output j is $\lambda_{i,j}$. Each input maintains one queue per output and $q_{i,j}$ denotes the backlog in the queue of input i for output j. This system is called a *virtual output buffer switch*.

In the basic model, one assumes that the switch operates in discrete time. That is, all the packets have the same size and take one unit of time to be transferred from the input buffer to the output link of the switch. The implementation of this switch requires that the variable size packets be decomposed in equal-size chunks that go through the switch and are reassembled at the output

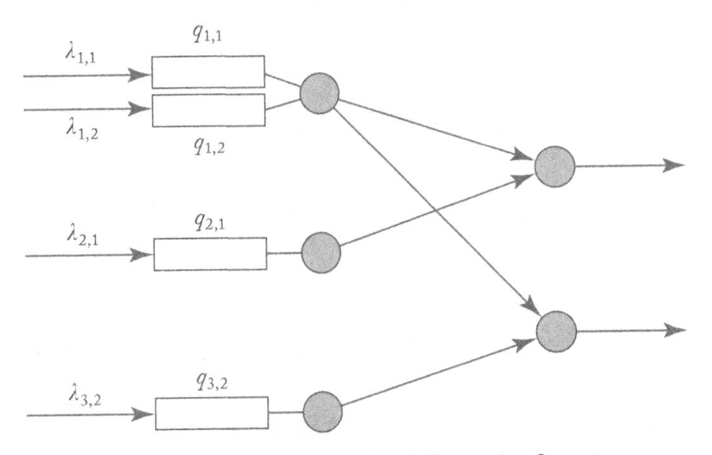

FIGURE 4.5: A switch with four active flows.

before being forwarded. This process incurs a cost of adding an header to each chunk and of waiting for the last chunk of a packet before forwarding it. The benefit is that the switch can transfer chunks in parallel without worrying about loss of synchronism.

In the example, the rates are *feasible* if they do not saturate an input or an output link. If we assume that these links have a unit rate, feasibility implies that

$$\sum_i \lambda_{i,j} < 1, \forall j \quad \text{and} \quad \sum_j \lambda_{i,j} < 1, \forall i. \tag{4.1}$$

Instability of Maximal Matching

A *maximal matching* is a matching of inputs to outputs that, at each time, matches the maximum possible number of non-empty queues; the maximal matching is chosen uniformly among all the maximal matchings. For instance, if all the queues of the switch in Figure 4.5 are non-empty, the following are the maximal matchings:

$$\{11, 32\}, \quad \{12, 21\}, \quad \text{and} \quad \{21, 32\}. \tag{4.2}$$

To be specific, assume that the arrivals are independent Bernoulli random variables in the different time slots, with the indicated mean values. That is, one packet arrives in a given time slot at queue (i, j) with probability $\lambda_{i,j}$, independently of the arrivals at other queues or in other time slots.

As shown in [McKeown et al. 1996], maximal matching may not be stable even if the rates are feasible. To see this, assume that the four rates are equal to $1/2 - \delta$ for some $\delta > 0$, so that (4.1) holds. Note that if packets arrive at both queues (2, 1) and (3, 2), which occurs with probability $(1/2 - \delta)^2$, then input 1 is served in at most two of the three maximal matchings (4.2), i.e., with probability 2/3. Hence, the service rate of input 1 is at most

$$\frac{2}{3} \times \left(\frac{1}{2} - \delta\right)^2 + 1 \times \left(1 - \left(\frac{1}{2} - \delta\right)^2\right).$$

Since the rate of arrivals at queue 1 is $1 - 2\delta$, we conclude that input 1 is unstable if

$$1 - 2\delta > \frac{2}{3} \times \left(\frac{1}{2} - \delta\right)^2 + \left(1 - \left(\frac{1}{2} - \delta\right)^2\right),$$

which occurs if $\delta < 0.0358$. Thus, maximal matching does not achieve the maximum throughput since the switch is not stable for some feasible rates when the switch uses that algorithm.

STABILITY OF MAXIMUM WEIGHTED MATCHING

A *maximum weighted matching* is a matching of inputs to outputs that maximizes the sum of the backlogs of the queues that are matched. For instance, if $q_{1,1} = 8$, $q_{1,2} = 8$, $q_{2,1} = 6$, and $q_{3,2} = 7$, the maximum matching is {11, 32} as it matches queues with a sum of backlogs equal to 15 whereas the other two maximal matchings corresponds to sums equal to 14 and 13, respectively.

To see that a maximum weighted matching is stable, as first shown in [Tassiulas and Ephremides 1992], one proves that

$$V(\mathbf{q}) = \frac{1}{2}\sum_{i,j} q_{i,j}^2$$

is a *Lyapunov function.* That is, one shows that

$$E[V(\mathbf{q}(n+1)) - V(\mathbf{q}(n)) \mid \mathbf{q}(n) = \mathbf{q}] \le -\alpha + \beta 1\{\mathbf{q} \in S\}, \tag{4.3}$$

where S is a finite set and $\alpha > 0$.

This inequality means that the function $V(\mathbf{q}(n))$ tends to decrease, on average, when $\mathbf{q}(n)$ is outside a finite set S. Since $V(\mathbf{q}(n)) \ge 0$, it cannot decrease all the time. Hence, it must be that $\mathbf{q}(n)$ spends a positive fraction of time inside the finite set S. This implies that the queue lengths are small for a positive fraction of the time and this proves the stability of the system, because the system must also be empty a positive fraction of the time.[3]

To show (4.3), we designate by $A_{i,j}(n)$ and $D_{i,j}(n)$ the number of arrivals and departures at queue (i, j) at time n and we note that

$$q_{i,j}(n+1) = \max\{q_{i,j}(n) + A_{i,j}(n) - D_{i,j}(n), 0\},$$

so that

$$\frac{1}{2}[q_{i,j}^2(n+1) - q_{i,j}^2(n)] = \le q_{i,j}(n)(A_{i,j}(n) - D_{i,j}(n)) + \frac{1}{2}(A_{i,j}(n) - D_{i,j}(n))^2.$$

Hence,

$$E\left[\frac{1}{2}[q_{i,j}^2(n+1) - q_{i,j}^2(n)] | \mathbf{q}(n) = \mathbf{q}\right] \le q_{i,j}[\lambda_{i,j} - D_{i,j}(n)] + K$$

for some constant K. Summing over i, j, we find

$$E[V(\mathbf{q}(n+1)) - V(\mathbf{q}(n)) \mid \mathbf{q}(n)] \le \sum_{i,j} q_{i,j}(n)[\lambda_{i,j} - D_{i,j}(n)] + B, \tag{4.4}$$

for some other constant B.

Now, if the rates are feasible, i.e., if they satisfy (4.1), there are some average service rates $d_{i,j}$ for the queues such that $\lambda_{i,j} \le d_{i,j} - \epsilon$ for all (i, j), for some $\epsilon > 0$. These service rates can be

3. One can show that when a Markov chain can go from every state to any other state, then either it spends a positive fraction of time in every state or a zero long-term fraction of time in every state.

implemented by using the different matchings some fractions of the time. This is possible because a doubly stochastic matrix that dominates $\{\lambda_{i,j}\}$ can always be written as a convex combination of permutation matrices that correspond to matchings. (This is the Birkhoff–Von Neumann theorem [Birkhoff 1946].) Moreover, the maximum matching chooses $\{D_{i,j}(n)\}$ to maximize

$$\sum_{i,j} q_{i,j}(n) D_{i,j}(n).$$

Hence,

$$\sum_{i,j} q_{i,j}(n) D_{i,j}(n) \geq \sum_{i,j} q_{i,j} d_{i,j}.$$

Substituting in (4.4), we find

$$\begin{aligned} E[V(\mathbf{q}(n+1)) - V(\mathbf{q}(n)) \mid \mathbf{q}(n)] &\leq \sum_{i,j} q_{i,j}(n)[\lambda_{i,j} - d_{i,j}] + B \\ &\leq -\epsilon \sum_{i,j} q_{i,j}(n) + B. \end{aligned}$$

This shows that (4.3) holds with $\alpha = 1$ and

$$S = \left\{ \mathbf{q} \mid \sum_{i,j} q_{i,j} \leq \frac{B+1}{\epsilon} \right\}.$$

This scheme favors the large queues. As a result, queues with little traffic tend to see larger delay. To mitigate this effect, one can use the delay of the packets at the head of the queues as a weight in the calculation of the maximum weight matching. See [Ji et al. 2013].

DELAY-BASED MATCHING

Instead of using the backlogs as the weights in the matching, one may also use the delays of the packets that are at the head of the line in each queue. The delay of a packet can be observed by adding a time-stamp to its header when it arrives at the switch.

To see why such a modification of the maximum weighted matching algorithm is stable, suppose that the delay of the head of the line packet in queue (i, j) is D_{ij}. If D_{ij} is large, then by the large of law numbers, the number of packets in the queue is $\lambda_{ij} D_{ij}$. Thus, using the delay as weight is equivalent to using q_{ij}/λ_{ij} as the weight. The stability of this scheme immediately follows by choosing $\sum_{i,j} q_{i,j}^2/\lambda_{i,j}$ as the Lyapunov function. This was originally proved in [Eryilmaz et al. 2005].

PRACTICAL ALGORITHM: ROUND-ROBIN

Calculating the maximum matching is NP-hard. Switches use simple heuristics instead. One of them is a round-robin grant and accept scheme [McKeown et al. 1993]. In this *round-robin matching*, the queues that are nonempty make a request to the corresponding outputs. The output grant requests in a round-robin order. Thus, if output 3 last granted input 1, its next grant will be in the order $(2, 3, \ldots, N, 1)$. Similarly, the inputs that receive multiple grants accept one in the round-robin order. This phase request-grant-accept is repeated a few times, each time removing the queues that have accepted a grant. Simulations show that this heuristic performs well in terms of delays and throughput. Many routers use a variation of this scheme.

Q-CSMA

Q-CSMA is a *randomized matching* algorithm based on an optimization problem. This mechanism works for variable size packets, which avoids the "cell tax" of the previous schemes. When using this scheme, the queues independently request outputs after a random time whose mean value decreases with the backlog in the queue. If the requested output is free, it accepts the transfer of one packet from the queue; otherwise, the input tries again later.

Thus, after finishing the transfer of a packet, input i generates independent random variables $\tau_{i,j}$ for all the outputs j for which $q_{i,j} > 0$. The mean value of $\tau_{i,j}$ decreases with $q_{i,j}$. If $\tau_{i,j}$ is the smallest of these random times, then input i requests queue j at that time. If output j is free, then input i transfers a packet from queue (i, j) to output j; otherwise, input i repeats the process. One can show that this scheme achieves the maximum throughput. As in the case of maximum weight, one may use the delays of the head of line packets instead of the backlogs. We analyze that scheme in the next section. The ideas are due to Libin Jiang (see [Jiang and Walrand 2010] for a more complete presentation).

4.5 RANDOMIZED MATCHING

In this section, we consider a generic problem of matching users and resources. The users need a set of resources to perform a task. The goal is to allocate the resources to users to maximize the utility of the tasks that they perform. We explain a randomized matching algorithm that achieves that objective.

PROBLEM FORMULATION

Consider N users who are competing for resources that they need to perform tasks. Specifically, there is a set $\mathcal{R} = \{1, 2, \ldots, R\}$ of resources and, to perform a task, user i needs the subset $\mathcal{R}_i \subset \mathcal{R}$ of resources for a random time that is exponentially distributed with rate μ_i. Thus, in this formulation, each user i requires precisely the set $\mathcal{R}_i$ of resources; there are no alternatives. The model can be extended to the case where different sets of resources enable user i to perform his task.

The subsets $\mathcal{R}_i$ are not disjoint, so the problem is to determine which user should get resources at any given time. The goal is to maximize the usefulness of the resources, measured by the utility that they provide the users.

We formulate the problem as follows:

$$\text{Maximize} \sum_{i=1}^{N} U_i(x_i)$$

$$\text{subject to feasibility.}$$

In this formulation, x_i is the rate at which user i is *working* on tasks and $U_i(x_i)$ is a concave increasing function. Note that user i *completes* tasks at rate $x_i \mu_i$ since each task completion requires μ_i^{-1} time units, on average. Thus, in our model, the utility is a function of the rate of work on tasks, as a mechanic being paid per hour instead of per completed task .

Since each resource j is used by at most one user at a time, we see that

$$\sum_{i=1}^{N} x_i 1\{j \in \mathcal{R}_i\} \leq 1, \quad \text{for } j = 1, \ldots, R.$$

These conditions characterize the *feasible* rates $\mathbf{x} = (x_1, \ldots, x_N)$.

At any given time, a set $z \subset \{1, \ldots, N\}$ of users have the resources they requested. We call z a matching. Thus, if $i \in z$, user i has the resources $\mathcal{R}_i$ under the matching z. A matching z is possible if

$$\mathcal{R}_i \cap \mathcal{R}_j = \emptyset \quad \text{if } i, j \in z,$$

because two users i and j in the matching z cannot use a common resource. Let $\mathcal{Z}$ denote the set of such possible matchings. If the rates $\mathbf{x}$ are feasible, they correspond to some probability distribution π on $\mathcal{Z}$ such that the matching $z \in \mathcal{Z}$ is used a fraction $\pi(z)$ of the time and

$$x_i \leq \sigma_i(\pi), \forall i$$

where

$$\sigma_i(\pi) := \sum_{z \in \mathcal{Z}} \pi(z) 1\{i \in z\}$$

is the fraction of time that user i has the resources $\mathcal{R}_i$. When these inequalities hold, we say that π achieves $\mathbf{x}$.

Algorithm 4.1 In principle, one could determine a randomized allocation scheme in a centralized way. One would solve the constrained optimization problem and determine the rates $\mathbf{x}^*$ that solve it. One would then find the probability distribution π^* on $\mathcal{Z}$ that achieves $\mathbf{x}^*$ and arrange to allocate the resources according to this distribution.

However, we are looking for a decentralized feedback scheme that adjusts $\mathbf{x}$ and π automatically, even when the rates μ_i are not known. We think of each user i placing requests at rate x_i in a counter that decreases at rate 1 whenever user i has the resources. Thus, these buffers maintain a count of the *deficit* of the users between their desired and actual resource possession times. Let q_i be the value of the counter of user i. The idea is then simple: one should increase the likelihood that user i gets the resources $\mathcal{R}_i$ when q_i gets large. The algorithm will do this automatically by having user i request the resources with an urgency that increases with q_i. Thus, the scheme is "the squeaky wheel gets the grease."

As before, we replace the optimization problem by that of maximizing the objective function minus a multiple of the drift of the Lyapunov function $V(\mathbf{q})$ where

$$V(\mathbf{q}) := \frac{1}{2} \sum_i q_i^2.$$

Note that

$$\frac{d}{dt} V(\mathbf{q}(t)) = \sum_i q_i(t)[x_i - \sigma_i(\pi)].$$

Accordingly, we want to maximize the difference between the utility and a multiple of the expression above, i.e.,

$$\sum_i U_i(x_i) - \rho \sum_{i=1}^{N} q_i[x_i - \sigma_i(\pi)]$$

subject to

$$\sum_{z \in \mathcal{Z}} \pi(z) = 1.$$

For the purpose of deriving our algorithm, we modify this problem as follows. The new goal is to maximize

$$\sum_i U_i(x_i) + \beta \left\{ H(\pi) - \sum_i \log(\mu_i)\sigma_i(\pi) \right\} - \rho \sum_{i=1}^{N} q_i[x_i - \sigma_i(\pi)] \tag{4.5}$$

subject to

$$\sum_{z \in \mathcal{Z}} \pi(z) = 1. \tag{4.6}$$

where β is a positive constant and

$$H(\pi) = - \sum_{z \in \mathcal{Z}} \pi(z) \log(\pi(z))$$

is the *entropy* of the allocation probabilities π. The additional term between curly brackets leads to an algorithm that can be implemented by a randomized request mechanism, as we explain below. Note that when β is small, the additional term is negligible because the term between curly brackets is bounded. Hence, the introduction of this term does not affect the objective function significantly.

To maximize (4.5) with respect to x_i, we find that

$$x_i \text{ maximizes } U_i(x_i) - \rho q_i x_i.$$

Thus, in a way similar to the congestion control scheme, the users maximize their net utility when they pay a price ρq_i for each new request that they place for the resource. One can view ρq_i as a *congestion price*.

Finally, one maximizes (4.5) with respect to π, subject to (4.6). We consider the Lagrangian $L(\pi, \lambda)$ given by

$$L(\pi, \lambda) = H(\pi) - \sum_i \sigma_i(\pi)\alpha_i - \lambda \left(\sum_{z \in \mathcal{Z}} \pi(z) - 1 \right),$$

where $\alpha_i = -\log(\mu_i) + \gamma q_i$, with $\gamma = \beta/\rho$. One finds that

$$\frac{\partial}{\partial \pi(z)} L(\pi, \lambda) = -1 - \log(\pi(z)) - \sum_{i \in z} \alpha_i - \lambda.$$

Setting that derivative to zero, we get

$$\pi(z) = A \exp\left\{ \sum_{i \in z} \alpha_i \right\} = A \Pi_{i \in z}, \mu_i^{-1} \exp\left\{ \gamma \sum_{i \in z} q_i \right\}, \tag{4.7}$$

where A is such that $\sum_z \pi(z) = 1$, i.e.,

$$A^{-1} = \sum_z \Pi_{i \in z} \mu_i^{-1} \exp\left\{ \gamma \sum_{i \in z} q_i \right\}.$$

The next step is to design an algorithm that implements these probabilities $\pi(z)$. One can verify that π is the invariant distribution of the continuous-time Markov chain on $\mathcal{Z}$ with transitions rates

$$Q(z, z \cup \{i\}) = \exp\{\gamma q_i\} \quad \text{and} \quad Q(z \cup \{i\}, z) = \mu_i. \tag{4.8}$$

Indeed, this Markov chain satisfies the following detailed balance equations:

$$\pi(z)Q(z, z \cup \{i\}) = \pi(z \cup \{i\})Q(z \cup \{i\}, z),$$

as can be verified using (4.7) and (4.8).

One implementation of this Markov chain is that user i requests the resources $\mathcal{R}_i$ with rate $\exp\{\gamma q_i\}$ where q_i is the value of the counter of user i. If all the resources $\mathcal{R}_i$ are available when user i requests them, he gets them and keeps them for an exponentially distributed random time with rate μ_i. If not all the resources $\mathcal{R}_i$ are available when user i requests them, he repeats the process after another random time.

It should be noted that the state transition diagram of the Markov chain is *insensitive*, so that the users get the resources with the fraction of times $\pi(i)$ even if the resource holding times are not exponentially distributed. Observe also that the invariant distribution assumes that the request rates remain constant, although they do not since the backlogs change as users place new requests and release the resource. Nevertheless, one can show that this scheme essentially achieves the maximum sum of utilities. (Technically, one shows that the Markov chain converges fast enough compared to the changes in backlogs.) A variation where the updates occur less and less frequently achieves the optimal utility.

Summarizing, the randomized matching algorithm is as follows:

- each user i maintains a counter q_i that tracks the deficit between the time during which he would like to perform tasks and the time during which he has the resources;
- user i places new requests in that counter with a rate x_i that maximizes $U_i(x_i) - \alpha q_i x_i$ where q_i is the value of the counter of user i and α is a positive constant; and
- when he does not have the resources, user i requests them with a Poisson rate equal to $\exp\{\gamma q_i\}$ where γ is a positive constant.

4.6 SUMMARY AND REFERENCES

Matching is a particular form of resource allocation where there is a set of M users and N resources. Each user gets at most one resource. The goal is to assign some of the resources to the users to achieve some utility of the resources.

In the housing and marriage matching problems, the goal is to reach a stable matching where no pair of users would rather swap their allocations. Although this goal is modest, we saw that in the marriage matching problem, there is no strategyproof scheme. These results are due to Gale, Roth, and Shapley [Roth 1984], [Roth 1988], and were implemented in applications that include matching medical school students to residency programs [Gale and Shapley 1962] and matching kidneys to patients [Roth et al. 2003].

In the switch scheduling problem, one matches inputs and outputs to achieve good performance measures. We saw that a suitable matching algorithm enables a virtual input buffer switch to mimic the behavior of an output buffer switch, with only a factor 2 speedup instead of a factor N that the latter design requires. To achieve stability under maximum throughput, we observed that maximal matching is not sufficient [McKeown et al. 1996]. Thus, being myopic and transferring the maximum possible number of cells across the switch at each time does not guarantee stability. Maximum weighted matching is stable, since it makes the sum of the squares of the queue lengths a Lyapunov function [Tassiulas and Ephremides 1992]. However, the complexity of that algorithm makes is impractical. We explained the randomized matching ideas based on Libin Jiang's work [Jiang and Walrand 2010] and we introduced a modification of the objective function to derive a robust algorithm that does not require knowing the mean service times.

CHAPTER 5
Collaboration

The Internet owes its existence to collaboration. Unlike the phone network that was built by monopolies in most countries, the Internet grew from the cooperation among thousands of autonomous domains that had very little in common other than joint adherence to protocols such as TCP/IP. The fact that this happened is all the more improbable given the fact that packets originating in one domain had to rely on the good will of other domains to reach their destination. More recently we have seen similar levels of networked cooperation in the design and implementation of Linux and in the creation of Wikipedia. What made this cooperation possible?

Economists have studied such situations in the form of games. Sometimes the parties cannot trust each other and that they have no way of enforcing an agreement. Thus, the only agreements that can be reached are self-enforcing, i.e., there is no incentive for any party to unilaterally deviate from the agreed-upon strategy. Economists have studied such scenarios in the form of *non-cooperative* games.

In other cases, there may be a way for the parties to enforce agreements made prior to finalizing their strategies. In such cases the games are *cooperative*. Cooperative game theory does not dwell on how inter-agent agreements are reached or even what these specific agreements are. It abstracts these processes as coalition payoffs. There is a real valued function, called the *characteristic function*, that specifies the aggregate payoff to the agents in any subset that agree to cooperate.

This chapter focuses on various models of cooperation, and explains important classical results. But as the reader will see, these results yield important insights but do not fully explain the cooperative behavior that established the Internet or Wikipedia.

5.1 PROFIT SHARING

Assume that the owner of a farm can generate a revenue equal to nA with n farm workers, for $n = 0, 1, 2$, where A is a given positive constant. Thus, if he hires one worker, together they generate a revenue A. If he hires two workers, together they generate a revenue $2A$. Assume that the owner hires two workers. How should he share the revenue $2A$ with the two workers?

This question may seem ill-formulated as many aspects of the situation are left unstated. Nevertheless, Shapley [Shapley 1988] proposes an answer to this question, based on some reasonable axioms about sharing rules. His answer is that the owner should get A and that each worker should get $A/2$.

As we saw in Section 1.2, Nash [Nash 1950] would choose the sharing to maximize the product of the changes in utility of the three agents. This solution requires specifying the utility function of the owner and the two workers, as well as their initial fortunes.

Another approach is suggested by Rubinstein [Rubinstein 1982]. He considers the situation where the owner enters into a bargaining process with the two workers. Assume that the negotiation plan is to meet every week to discuss the fraction x that should go to the owner. The value of the agreement shrinks by a factor β every week. The first week, the owner insists in keeping a fraction x. The second week, the workers can come back with a counter-proposal where they offer a fraction y to the owner, and the bargaining can continue in that way until an agreement is reached. It can be shown that the rational choice in this situation is for everyone to agree that $x = 1/(1+\beta)$ on the first day of the bargaining.

Thus, Shapley, Nash, and Rubinstein propose three different approaches to determine how collaborating users should share the profit of their collaboration. These approaches correspond to different contexts. One should keep in mind that even the formulation of the objective is a nontrivial question, as was also the case for resource allocation problems.

Example 5.1 (Why are the Berkeley Streets so Bad?) The following simple example (borrowed from [Hauert 2005]) illustrates the difficulty in financing public goods.

The citizens of Berkeley are asked to contribute to improvements to the notoriously bad city streets. The aggregate increase in utility to the N citizens is Ax if they collectively invest x. The constant A is larger than 1. All the citizens share equally in the increase.

Thus, if every citizen contributes an amount y, the increase in utility is ANy and each citizen gets a net increase in utility $(ANy)/N - y = (A-1)y$. Since $A > 1$, it seems rational for every citizen to invest as much as he can afford. (To simplify our story, we ignore the possible existence of other options that bring a higher return.) However, every citizen realizes that if the others collectively invest z, by investing y he gets a net revenue

$$\frac{A(z+y)}{N} - y = \frac{Az}{N} + y\left(\frac{A}{N} - 1\right).$$

Indeed, the return Ay or the investment y of each citizen is shared among the N citizens. Thus, if $A < N$, it is a dominant strategy for that citizen to invest $y = 0$. As a result, no one invests in street improvements.

The example shows that users might all benefit by collaborating but, selfishly, they each have an incentive to deviate from the collaboration. This is an example of what is called a "Prisoner's Dilemma."

INVESTING IN BROADBAND

The N residents of an apartment building contemplate getting together to invest in broadband access. If n residents are to be connected, the cost per resident is $c(n)$ where $c(n)$ is decreasing in n. For $i = 1, \ldots, N$, the value of the connection to resident i is v_i, which is known only to that resident.

The goal is to design a mechanism to divide fairly the cost among the residents. A simple scheme is to start by assuming that the N residents will participate and announce a price $c(N)$. The residents whose valuation v_i is less than $c(N)$ will then drop out, and a subset of N_1 residents will remain interested. One then announces a price $c(N_1)$. Another subset of residents may drop out because $c(N_1) > c(N)$. One continues in this way until no more residents drop out and one ends up with a set, say N_k (possibly empty) of citizens willing to pay $c(N_k)$.

This scheme, is an example of a *carving* algorithm, and is *budget-balanced* since the money collected covers exactly the cost. This mechanism is also called the Moulin Mechanism after [Moulin and Shenker 2001]. Moreover, it is easy to see that it is *group strategyproof*. That is, no subgroup of residents have an incentive to collude and lie about their true valuation. Indeed, if a group of citizens pretend that their valuation is less than an announced price, they drop out and get no utility. Residents cannot reduce the price they have to pay by pretending that their valuation is smaller than it actually is.

5.2 COST SHARING AND CROSS MONOTONIC MECHANISMS

In the previous example it made sense to apportion the cost equally among the subset of residents who subscribe to the service. However, in the more general case this may not be true. For example, suppose the agents are resident at the nodes of a multicast tree as in Figure 5.1. The actual costs are derived from the link costs marked in black. If only users 3 and 5 opt for service, then the argument could be made that 5 should be apportioned more of the total cost than is 3 since it costs more to reach him.

The cost $C(S)$ of providing service to a subset S of the set N of users is the sum of the cost of the links needed to connect these users to the root. For instance, $C(\{1\}) = 4$, $C(\{2\}) = 5$ and $C(\{1, 2\}) = 8$. Because users share links, we see that

$$C(S \cup T) \leq C(S) + C(T) - C(S \cap T)$$

for all subsets S, T.

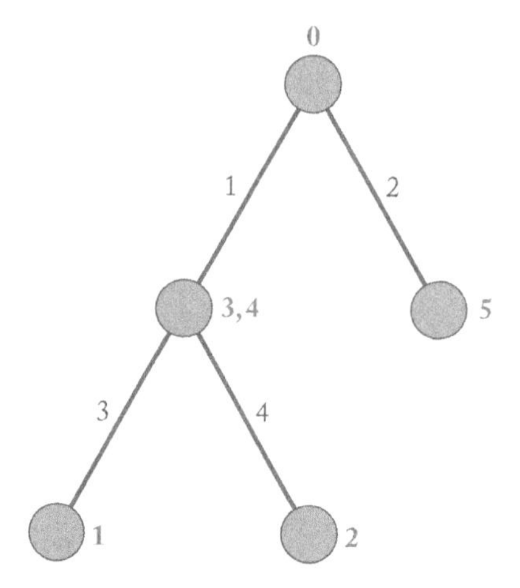

FIGURE 5.1: Multicast Cost Sharing. In the figure, 0 is a gateway to the Internet and five residents $1, \ldots, 5$ are connected to the gateway by the tree that has four edges.

This is an example of a *submodular* cost function. Submodular cost functions exhibit economies of scale: adding more users reduces the cost per user.

Given the cost function C, how should one allocate the costs for any subset of the users S that choose to participate? There are clearly many ways to do this, but one logical scheme is to have the users share equally in the cost of each link they use to connect to the root. This is an example of a *cost allocation scheme*. Continuing with the example in Figure 5.1, let $c_i(S)$ be the cost assigned to user i when the users in set S are connected. Then $c_1(\{1, 2, 3, 4, 5\}) = 3 + 1/4$, since user 1 alone pays for the link with cost 3 and he pays for one fourth of the cost of link with cost 1.

The cost sharing scheme of our example is perfectly *budget balanced* since all the costs are perfectly recovered. But some cost sharing schemes may not have this property. We say that a cost sharing scheme is α-budget balanced if the cost recovered is always within a factor α of the cost incurred. For example, suppose $C(S) = \sqrt{|S|}$ and the scheme allocates only $\frac{1}{|S|}$ to each participating user i. Then the cost recovered is always $|S| \times \frac{1}{|S|} = 1$ but the cost incurred is $\sqrt{|S|}$. If there are n users in all, this cost sharing scheme is $\frac{1}{\sqrt{n}}$-budget balanced.

Next, let us consider how to implement a particular cost sharing scheme, η. If each user i values the service at v_i dollars and the set of participating users is S, it must be true that $v_i \geq c_i^{\eta}(S)$. Otherwise user i is participating at a loss. This requirement is called **Voluntary Participation**. Also, it might be the case that the set of all users can collude and attempt to lie about their private valuations, v_i,[1] and a group strategyproof implementation is thus desirable. Finally, as much of the costs should be allocated to the participating users as in the cost allocation scheme.

1. They cannot, however, transfer money among themselves

It turns out that when one restricts oneself to a particular type of cost allocation scheme, it is possible to jointly meet these goals. The cost allocation scheme η must be *cross monotonic*, i.e., for any user i and $S \subset T$:

$$c_i^{\eta}(S) \geq c_i^{\eta}(T).$$

As more users are connected the allocated cost to each user can only go down.

In the shared multicast tree example, the proposed allocation scheme is cross-monotonic. Following [Moulin and Shenker 2001], the carving algorithm applied to such a problem yields a budget balanced, group strategyproof with Voluntary Participation. In particular, assume that the values for the users are $v_i = i$, for $i = 1, \ldots, 5$. The carving process starts with the set $N = \{1, 2, 3, 4, 5\}$. Since $c_1(N) = 3.25 > v_1 = 1$ and $c_2(N) = 4.25 > v_2 = 2$, users 1 and 2 drop out. Then, $N_1 = \{3, 4, 5\}$. Now, $c_3(N_1) = 0.5 < v_3 = 3$, $c_4(N_1) = 0.5 < v_4 = 4$ and $c_5(N_1) = 2 < v_5 = 5$, so that the carving process terminates.

This can be generalized to get the following result.

Theorem 5.1 For any cross-monotonic, α-budget balanced cost sharing scheme, the carving mechanism is always group strategyproof, α-budget balanced and obeys voluntary participation.

The converse of the theorem, i.e., given a group strategyproof mechanism one can always construct a cross monotonic cost sharing scheme, only holds for the case that the cost function is submodular.

Another property (which is trivially satisfied) of the carving mechanism is that given any set of values v_j, $j \in N - \{i\}$, there is a value v_i^+ that ensures i of service. Thus, this a property of *non-exclusion*.

Finally, let the efficiency of a mechanism followed by N agents be

$$\sum_{1=1}^{N} \max\{0, v_i - c_i(S)\},$$

where S is the subset of the users who choose to participate. The carving mechanism can in fact be quite inefficient, an issue that we will take up further in the next section.

5.3 SHAPLEY VALUE

When a group of agents decide to collaborate, they may not all bring the same value to the table. Shouldn't they divide the value derived from the cooperation in proportion to their respective contribution? But how can one compute that value?

Let $v(S)$ be value when a subset S of the set N of agents cooperate. Let us assume that the function $v(\cdot)$ is *superadditive*, i.e., such that, for any $S, T \subset N$ with $S \cup T = \emptyset$, one has

$$v(S \cup T) \geq v(S) + v(T). \tag{5.1}$$

This condition gives the agents an incentive to cooperate.

Given the function $v(\cdot)$, one wishes to assign a value $\phi_i(v)$ to each individual agent $i \in N$. Shapley proposed the following reasonable properties of the assignments $\phi_i(v)$:

S1 Efficiency. $\sum_{i \in N} \phi_i(v) = v(N)$, i.e., the values of the agents sum up to the total value of all agents co-operating.

S2 Symmetry. Suppose there are two agents i, j such that for any coalition S that does not include either of them one has $v(S \cup \{i\}) = v(S \cup \{j\})$. Then $\phi_i(v) = \phi_j(v)$.

S3 Additivity. Suppose two scenarios are defined for the agents in N with value functions v and w. If they are combined into a single scenario x such that the $x(S) = v(S) + w(S)$. Then

$$\phi_i(x) = \phi_i(v) + \phi_i(w).$$

S4 Dummy. If an agent i does not contribute anything, i.e., if $v(S \cup \{i\}) = v(S)$ for all coalitions, then $\phi_i(v) = 0$.

SHAPLEY VALUE

Surprisingly, it turns out that there is exactly one form of $\phi(v)$ that satisfies S1-S4. This is called the *Shapley value*.

Theorem 5.2 In a cooperative game with n players in a set N and a real valued function v defined on all subsets of N that satisfies (5.1), the only real valued functions $\{\phi_i(v), i \in N\}$ that satisfy S1-S4 are given by:

$$\phi_i(v) = \frac{1}{n!} \sum_{\pi \in \Pi} [v(p_\pi^i \cup \{i\}) - v(p_\pi^i)], \tag{5.2}$$

where Π is the set of all permutations of N, and for $\pi \in \Pi$, $p_\pi^i = \{j : \pi(i) > \pi(j)\}$,[2] i.e. p_π^i is the subset of N which consists of all the agents who precede i in permutation π.

Proof. It is immediate to verify that the value function satisfies the properties $S1$–$S4$. The only thing to prove is the uniqueness. Let $\mathcal{S}$ be the collection of non-empty subsets of N.

One proves uniqueness by observing that any superadditive value $\{v(S), S \in \mathcal{S}\}$ can be written uniquely as

$$v(S) = \sum_{T \in \mathcal{S}} a_T v_T(S), \forall S \in \mathcal{S}, \tag{5.3}$$

2. If $N = \{1, 2, 3, 4\}$, then for the permutation $\pi = (2, 1, 3, 4)$, $p_\pi^1 = \{2\}$, $p_\pi^2 = \emptyset$, $p_\pi^3 = \{2, 1\}$ and $p_\pi^4 = \{2, 1, 3\}$.

for some constants $\{a_T, T \in \mathcal{S}\}$ where

$$v_T(S) = 1\{T \subset S\}, \forall S \subset N.$$

The values $v_T(\cdot)$ correspond to *unanimity games* where the value of a coalition S is one if S contains T and zero otherwise.

To see why this is the case, note that there are $2^n - 1$ functions $v_T(\cdot)$ for $T \in \mathcal{S}$.[3] This is the dimension of the domain of the values $v(\cdot)$ since $v(\emptyset) = 0$. Thus, it suffices to show that these functions $v_T(\cdot)$ are linearly independent. That is, one has to show that if

$$0 = \sum a_T v_T(S) = 0, \forall S \subset T, \tag{5.4}$$

then $a_T = 0, \forall T \in \mathcal{S}$. Assume as a contradiction hypothesis that this is not the case and let T_0 be the set with the smallest cardinality such that $a_{T_0} \neq 0$. Then,

$$v_T(T_0) = 0, \forall T \neq T_0.$$

Indeed, if $T \neq T_0$, then it is not possible that $T \subset T_0$, since $T \subset T_0$ and $T \neq T_0$ implies that T has fewer elements than T_0, which is not possible. Hence,

$$\sum_{T \in \mathcal{S}} a_T v_T(T_0) = a_{T_0}.$$

Thus, by (5.4), $a_{T_0} = 0$, a contradiction.

Now, if one defines an allocation for every basis value $v_T(\cdot)$, then the allocation is determined from (5.3) by the axiom of additivity. Moreover, the axioms of efficiency, symmetry and dummy player imply that the allocation for the basis value $v_T(\cdot)$ is given by $\{\psi_T(1), \ldots, \psi_T(n)\}$ where

$$\psi_T(i) = \frac{1}{|T|} 1\{i \in T\}.$$

Hence, there is only one allocation that satisfies Shapley's axioms and we know that the Shapley value does. ∎

The Shapley value itself has the following intuitive interpretation. Each permutation π corresponds to an order in which the agents decide to cooperate. p_π^i is the subset of N which consists of all the agents who arrived before i in that permutation. We consider the marginal contribution of i to this set, i.e., how much does the cooperation of i add to the value of the coalition p_π^i.

3. Recall that $n = |N|$.

TABLE 5.1: The agents arrive in a random order and contribute to the revenue.

Order			Contribution of		
			O	$W1$	$W2$
O	$W1$	$W2$	0	100	100
$W2$	O	$W1$	100	100	0
$W1$	$W2$	O	200	0	0
O	$W2$	$W1$	0	100	100
$W1$	O	$W2$	100	0	100
$W2$	$W1$	O	200	0	0

These marginal contributions are averaged over the $n!$ permutations to yield ϕ_i. In calculations, the following form is more convenient:

$$\phi_i(v) = \sum_{S \subset N - \{i\}} \left[\frac{|S|!(n-1-|S|)!}{n!} v(S \cup \{i\}) - v(S) \right]. \tag{5.5}$$

Example 5.2 (Profit Sharing) Assume that a farm owner cannot produce any revenue without workers. With one worker, he can produce \$100.00 and with two workers he can produce \$200.00. Say that the owner hires the two workers. How should the owner and the two workers share the \$200.00 revenue? Table 5.1 shows the six permutations of the agents and how much each contributes in each case. For instance, if they arrive in the order O, $W1$, $W2$, then the owner does not contribute to the revenue because his arrival does no increase the revenue; $W1$ adds \$100.00 to the revenue because before his arrival the owner could not produce anything whereas with $W1$ he can produce \$100.00; also, $W2$ adds \$100.00 because with him they can produce \$200.00 instead of \$100.00. The cases that correspond to other orders have a similar justification. The average contribution of the owner is then $(0 + 100 + 200 + 0 + 100 + 200)/6 = 100$. Similarly, one finds that the average contribution of $W1$ and $W2$ is 50. According to Shapley, the owner should get \$100.00 and each worker should get \$50.00.

Example 5.3 (Profit Sharing (continued)) Assume now that the farm owner can hire n workers under the same conditions as the previous example, i.e., with n workers he can produce $\$100n$ in revenue. Consider any of the $n!$ permutations of the n workers. When the owner is the kth position of this permutation ($k = 1, \ldots, n+1$), his marginal contribution is $100(k-1)$. Thus,

$\phi_o(v) = \frac{100}{n+1}(1+2+\cdots+n) = 50n$. By symmetry all the workers must have the same Shapley value, ϕ_w. By efficiency, $\phi_o + n\phi_w = 100n$. Thus, $\phi_w = 50$.

Example 5.4 (Security Council Voting Power [Roth 1988]) The Shapley Value can also help to reveal the relative power of different classes of voters. Suppose there are 15 voters, but 5 special voters have veto power. A resolution is only carried when at least 4 of the remaining 10 voters vote in the affirmative along with the special voters. Suppose we want to find the relative power or importance of these two classes of voters. Let $v(S) = 1$ if S satisfies the conditions for carrying a resolution and $v(S) = 0$ otherwise. By symmetry all five of the special voters have the same Shapley Value, ϕ_s as do the regular voters ϕ_r. By efficiency, $10\phi_r + 5\phi_s = 1$. Now, for a regular member to make a positive contribution, all five of the special members and three of the other nine regular voters must have already be part of the coalition. There are $\binom{9}{3}$ ways to choose the three regular voters and using (5.5) for each of these choices for which $|S| = 8$, there are $8!(15-1-8)! = 8!6!$ different coalitions. Thus there are a total of $\binom{9}{3}8!6!$ such coalitions. Dividing by 15! we get $\phi_r(v) = 0.00186$. By efficiency, $\phi_s = 0.196$ so that a special voter is about 105 times more powerful than a regular voter.

Example 5.5 (Cost Sharing) We want to share the cost of a service among n agents in the set N, where the ith agent values the service at $v_i \geq 0$ dollars. Let $C(S)$ be the cost to deploy the service to an arbitrary subset S of agents. Then

$$v(S) = \sum_{i \in S} v_i - C(S).$$

We have already studied this problem in Section 5.2 where we show that the cross-monotonic mechanisms are attractive. As we did there, assume the cost is submodular. Consequently, for all $S \subset T \subset N$:

$$C(T) \leq C(S) + C(T \setminus S).$$

Then $v(S)$ obeys (5.1). Also,

$$v(S \cup \{i\}) - v(S) = v_i - C(S \cup \{i\}) + C(S).$$

From (5.5):

$$\phi_i(v) = \sum_{S \subset N - \{i\}} \frac{|S|!(n-1-|S|)!}{n!}(v_i - C(S \cup \{i\}) + C(S)). \tag{5.6}$$

The part of ϕ_i that depends on the cost function is:

$$c_i^{sh}(S) = \sum_{S \subset N - \{i\}} \frac{|S|!(n-1-|S|)!}{n!} C(S \cup \{i\}) - C(S)),$$

where, $c_i^{sh}(S)$ is a cost apportionment method, i.e., a method to divide the cost among the agents. Comparing to (5.5), we note that $c_i^{sh}(S)$ obeys S1–S4 and therefore we refer to this as the Shapley cost apportionment method.

SHAPLEY VALUE AND MULTICAST

It is interesting to observe that the cost apportionment method described in Figure 5.1 on Page 98 is identical to the Shapley cost apportionment method applied to the cost function of the multicast problem! To see that this is true, observe that this allocation meets S1-S4, and since $-C(S)$ follows (5.1), the c_i are the unique Shapley values.

It is easy to see that in general, if $C(S)$ is submodular then the cost apportionment method c^{sh} must be cross-monotonic. To see this, suppose we start from a set S and add a new agent a, and consider how c_i^{sh} changes for some agent $i \in S$. Now

$$
\begin{aligned}
C(S \cup \{a\} \cup \{i\}) - C(S \cup \{a\}) &\leq C(S \cup \{i\}) - (C(S \cup \{a\}) - C(\{a\})) \\
&\leq C(S \cup \{i\}) - C(S). \qquad (5.7)
\end{aligned}
$$

Then, from (5.2) it is clear that the $c_i^{sh}(S \cup \{a\}) \leq c_i^{sh}(S)$. Also if $C(S)$ is strictly submodular (i.e., the inequality condition is strict), then c_i^{sh} is strictly monotonic.

EFFICIENCY OF SHAPLEY VALUE

We mentioned that cross-monotonic mechanisms can be quite inefficient. For a given $v = (v_1, \ldots, v_n)$ and cross-monotonic apportionment method η, let $\gamma(\eta, v)$ be the efficiency loss due to η when v is the profile of utilities. I.e., if the optimal welfare possible is $W(v)$ and the mechanism yields $\hat{W}(v)$, then

$$\gamma(\eta, v) = W(v) - \hat{W}(v).$$

Then we can define the worst case efficiency loss due to η as

$$\gamma(\eta) = \sup_{v} \gamma(\eta, v).$$

The following theorem is shown in Moulin and Shenker (2001):

Theorem 5.3 Among all cost apportionment algorithms η derived from cross-monotonic mechanisms, the Shapley cost apportionment method has the uniquely smallest efficiency loss which is given by:

$$\gamma(\eta^{sh}) = \sum_{S \subseteq N} \left(\frac{|S|!(n - |S|)!}{n!} C(S) \right) - C(N).$$

Stability of Shapley Value

We have seen that if a value is superadditive, there is only one allocation that satisfies the axioms S1–S4 and it is the Shapley value.

However, is this allocation in the core? That is, as we explained in Section 4.1, is it the case that no subset of the agents would be better off by splitting from the rest of the agents?

In general, the answer is negative. As an example, consider a vote that requires a strict majority to approve a spending bill of one million dollars. There are four political parties that can participate in the vote and split the million dollars if it is voted. The parties are A, B, C, D and they have 45, 25, 15, and 15 representatives, respectively. Thus, if $\{B, C, D\}$ are allocated less than the full amount, they will leave the group and form their own coalition. However, if they get the full amount, A can form a coalition with the party among $\{B, C, D\}$ that gets the smallest amount. For instance, say that A, B, C, D get 0, 0.4, 0.3, 0.3, respectively, then A can form a coalition with C and agree to get 0.6 and 0.4, respectively. Thus, the Shapley value that allocates a positive amount to each agent is not stable: it is not in the core. In fact, there is no allocation in the core for this example.

Recall that a value $v(\cdot)$ is *supermodular* if

$$v(S \cup T) \geq v(S) + V(T) - v(S \cap T), \forall S, T \subset N. \tag{5.8}$$

Equivalently, the function is supermodular if

$$v(A \cup B \cup C) - V(A \cup B) \geq V(B \cup C) - V(B),$$

for all disjoint $A, B, C \subset N$.[4] Thus, the value increases more by adding C to $A \cup B$ than by adding it to B. This is a form of increasing return, or convexity.

One has the following theorem.

Theorem 5.4 If the value is supermodular, then the Shapley value is in the core.

Proof. Fix an order of the players: $\pi = (1, 2, \ldots, n)$. Define the allocation

$$\phi_\pi(i) = v(\{1, , 2 \ldots, i\}) - v(\{1, 2, \ldots, i-1\}).$$

This allocation is efficient (it allocates the full value) and individually rational since the allocation of every user is nonnegative. Also, the supermodularity implies that this allocation is in the core. Indeed, consider a subset $S = \{a_1, a_2, \ldots, a_k\}$ of the agents with $1 \leq a_1 < a_2 \ldots < a_k \leq n$. By supermodularity, for $l = 2, \ldots, k$, one has

$$\begin{aligned} &v(\{a_1, a_2, \ldots, a_l\}) - v(\{a_1, \ldots, a_{l-1}\}) \\ &\leq v(\{1, 2, \ldots, a_l\}) - v(\{1, 2, \ldots, a_l - 1\}) = \phi_\pi(a_l). \end{aligned}$$

4. To see the equivalence, let $A = S \setminus T$, $B = S \cap T$ and $C = T \setminus S$.

Summing this expression over l gives

$$v(S) \leq \sum_{l=1}^{k} \phi_{\pi}(a_l).$$

Hence, the set S of agents is better off by staying in the group than by splitting off.

Thus, each ordering π corresponds to an allocation that is in the core. The Shapley value is the average of these allocations over all the orderings and is therefore also in the core. ■

5.4 JOINT PROJECT

This section explores some aspects of projects where a client solicit the collaboration of participants through a crowd-sourcing application, such as Amazon's Mechanical Turk. Some projects require sufficiently many participants to be successful. Others require only one participant. The web enables a market where clients and participants can meet. The basic law of supply and demand suggests that these crowd-sourcing applications may result in a severe competition among participants, which may result is very low wages. Our models confirm this prediction. One may hope that, as these systems mature, the demand will increase and correct this unfortunate situation.

Imagine a project that requires the collaboration of multiple participants. This project can be collecting data about traffic, effectiveness of medical treatments, or a marketing survey. The value of the collaboration increases with the number of participants. To incentivize collaboration, the initiator whom we call the *client* proposes some compensation. The problem we explore is the design of this compensation. We study a few possible models.

Model 5.1 (Fixed Reward) The client proposes a reward R for each collaborator. How does he optimize R? Assume there are N potential collaborators. Each potential collaborator, called *agent*, n has a random cost C_n. The costs are i.i.d. with cumulative probability distribution $F(\cdot)$. We assume that N and $F(\cdot)$ are known to everyone. The client has a value $V(n)$ for the collaboration of n agents.

In this model, we assume that the agents are price-takers. That is, agent n collaborates if $C_n \leq R$, which occurs with probability $F(R)$. Thus, the number Z of collaborators is binomial with parameters N and $F(R)$. The expected net value of the collaboration for the client is then

$$E(V(Z) - RZ) = E(V(Z)) - RNF(R).$$

In principle, one can find the optimal value of R, at least numerically.

Model 5.2 (Gradual Increase of Reward) In this model, the client increases the reward R slowly. The agents are price takers and accept to collaborate as soon as R exceeds their cost. Thus, if $Z(R)$

is the number of clients with a cost less than or equal to R, the reward of the client when the reward is R is equal to

$$V(Z(R)) - Z(R)R.$$

The problem for the client is to decide when to stop.

Model 5.3 (Strategic Agents) This model is similar to Model 2, except that the agents are strategic and independently decide when to accept to collaborate.

There are 2 agents and the client proposes a reward k at step k, for $k = 1, 2, \ldots$. If the two agents accept to collaborate at step k, each one is selected with probability $1/2$ and the game stops. If only one agent accepts at step n, he gets the reward and the game stops. Assume that the game lasts at most T steps and that $T \geq 2$. This is a version of the *centipede* game [Rosenthal 1981]. The surprising result is that rational agents should accept at step 1 and get an expected reward equal to $1/2$, although they know that the expected reward would be $T/2$ if they held off until step T.

We first review the notion of equilibrium for a dynamic game such as this one. A player strategy specifies the probabilities of the different possible actions for any possible history of the game so far. A *subgame perfect equilibrium* (SPE) is a pair of strategies, one for each player, such that no player can gain by deviating from his strategy, for any possible history of the game. Here, the strategies specify the probability $p(1, k)$ and $p(2, k)$ that agent 1 and 2, respectively, accept the offer at each step k of the game. A symmetric SPE is such that $p(1, k) = p(2, k) = p(k)$ for all k.

Theorem 5.5 The strategies $p(k) = 1$ for all k are a symmetric SPE.

Proof. This game is analyzed by backward induction. Let $V(k)$ be the value of the game, i.e., the maximum expected reward for an agent, if no agent has accepted prior to time k, for $k = 1, \ldots, T$. Note that $V(T) = T/2$ since at the last step both agents should accept.

Assume that no agent has accepted before time k and that agent 2 accepts with probability $p \in (0, 1)$. If agent 1 accepts at time k, his reward is k with probability $1 - p$ (if the other agent does not accept) and it is k with probability $p/2$ (if the other agent accepts and the client selects agent 1). We denote this expected reward by $R(1, p)$. Thus,

$$R(1, p) = k(1 - p) + \frac{k}{2}p.$$

If agent 1 does not accept, then with probability p his reward will be zero and with probability $1 - p$ his maximum reward will be $V(k + 1)$. Thus, his expected reward is

$$R(0, p) = (1 - p)V(k + 1).$$

Assume, as an induction hypothesis, that $V(k+1) = (k+1)/2$. Then we find that

$$R(1, p) = k(1-p) + \frac{k}{2}p \geq R(0, p) = (1-p)\frac{k+1}{2}.$$

Thus, under this assumption, agent 1 is always better off by accepting at time k. Thus, a symmetric equilibrium is such that both agents choose $p(k) = 1$. Also, with this choice, $V(k) = k/2$, so that the induction hypothesis also holds at step k. Since $V(T) = T/2$, we conclude that $p(k) = 1$ for all k is indeed a SPE for this game. ■

5.5 CONTRACTS

In this section we review the design of contracts. The basic problem is for a contractor to design contracts that he proposes to subcontractors. A contract specifies a quantity of work for some remuneration. The contractor defines a collections of contracts and each subcontractor selects the contract that is most advantageous for him. If the contracts are well designed, the selection is also most advantageous for the contractor. In some situations, the contractor does not know the characteristics of the subcontractors, such as their efficiency. We discuss an example from [Duan et al. 2013].

Model 5.4 (Linear Production) There are N potential collaborators, called agents. Each agent i faces a cost $c_i > 0$ and produces a quantity of work $q_i > 0$ per unit of effort. Assume that agent i spends $w_i \leq b_i$ units of effort, for $i = 1, \ldots, N$. Here, b_i is an upper bound on the quantity of effort that agent i can produce. Together, the agents produce a quantity x of work, where

$$x = \sum_{i=1}^{N} q_i w_i.$$

We assume that the utility of this work is $U(q)$ to the client who contracts out the work, where $U(\cdot)$ is an increasing concave function and q is the quantity of effort that the agent produces.

We consider two situations. In the *full information* case, the client know the characteristics of all the agents. In the *incomplete information* case, he does not.

FULL INFORMATION

If the client knew the characteristics (c_i, q_i, b_i) of all the agents, he could sign separate contracts whereby agent i would be paid $c_i w_i$ to provide w_i units of effort. The client would then derive a net surplus S where

$$S = U\left(\sum_{i=1}^{N} q_i w_i\right) - \sum_{i=1}^{N} c_i w_i.$$

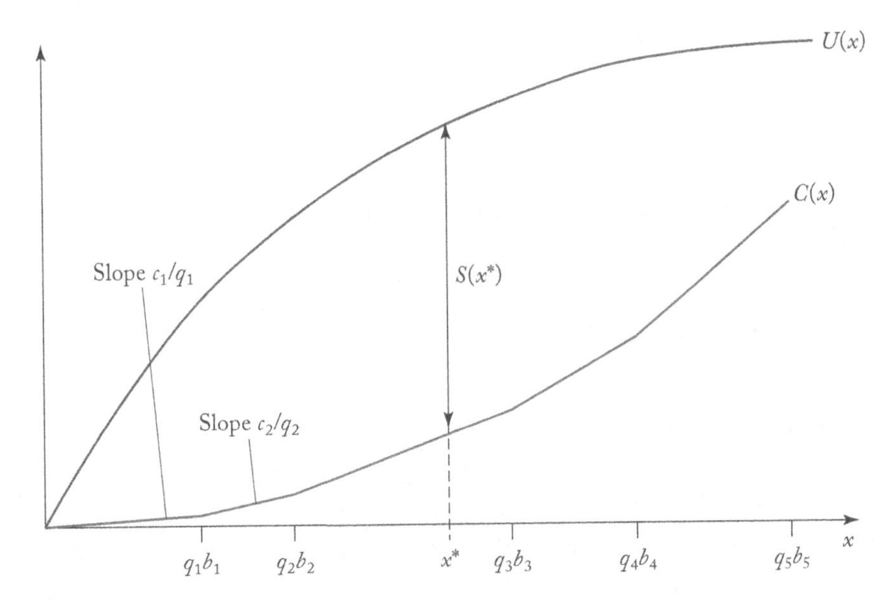

FIGURE 5.2: The utility $U(\cdot)$ and cost $C(\cdot)$ functions.

In this situation, the client faces the problem of maximizing S over $(w_1, \ldots, w_N)$ subject to $0 \leq w_i \leq b_i$ for $i = 1, \ldots, N$.

The solution of the problem is simple. One first ranks the agents in increasing order of c_i/q_i. Assume then that the order is $(1, 2, \ldots, N)$. One then considers the functions shown in Figure 5.2. The function $C(x)$ is the minimum cost to produce a quantity x of work. From the figure, one finds the value x^* of x that maximizes $U(x) - C(x)$. One then reads off the amount of work the client should ask from every agent, and consequently the price he will pay them. In the figure, agents 1 and 2 produce their maximum amount of effort, agent 3 produces some effort, and the other agents do not collaborate.

For instance, in the case of a large number of agents, one might be able to estimate the number of agents of the different types, so that this case applies to that situation.

INCOMPLETE INFORMATION

In many situations, it is more realistic to assume that the client knows the type of agents that are potential collaborators but does not know the type of every agent and may not be able to offer individual contracts.

Assume that there is only one user of each type and that the values of c_i/q_i are all distinct. In this case, say that the client offers a fixed price c_i/q_i per unit of work. Agents $\{1, \ldots, i\}$ accept the contract and the others do not. Consequently, they will produce the quantity of work $b_1 + \cdots + b_i$

for a price $(c_i/q_i)(b_1 + \cdots + b_i)$. The surplus of the client is then

$$S_i = U(b_1 + \cdots + b_i) - \frac{c_i}{q_i}(b_1 + \cdots + b_i).$$

The client chooses the value of i that maximizes that expression, and this defines the contract.

One can verify that the lack of information and the inability to write individual contracts results in a loss of surplus for the client.

5.6 SUMMARY AND REFERENCES

In contrast with the previous chapters, we focused on cooperation among users. The basic model is that by collaborating, users can produce some value. The question is then how they should share the profit of their collaboration and how to design a mechanism that encourages such joint effort.

We started by reviewing the ideas of Shapley [Roth 1988], Nash [Nash 1950], and Rubinstein [Rubinstein 1982] who propose different methods for sharing the profit that a collaboration generates. We then showed through an example that public investments may correspond to a prisoner's dilemma where individual citizens have an incentive not to collaborate. Such results have important consequences in political economics.

We explained the cross-monotonic mechanisms for the allocation of the cost of a project with many beneficiaries in the case of investments in networks.

In Section 5.4, we considered problems that are motivated by "crowdsourcing" applications. Simple examples show that such applications generate a global competition of workers that leads them to accept very low wages. The fact that these workers do not know one another preclude collective bargaining and may possibly set back the cause of workers.

Finally, we explained some simple ideas of contract theory. See [Bolton and Dewatripont 2005] for a detailed presentation of this important topic.

CHAPTER 6
Stability

In Chapter 2, we studied the allocation of bandwidth to different users. Specifically, we examined the rates of the flows that maximize the total utility of the users. The constraint was that the total rate of flow on each link should not exceed the transmission rate, or capacity, of the link.

This formulation abstracts away the discrete nature of the packets that the Internet transmits. It uses the intuitive notion that the average rates of packet flows suffice to determine the feasible operating points of the network. Technically, this condition on the average rates of flows should guarantee that the network is stable, i.e., that the queues do not grow without bounds and that all the packets make it across the network after a finite time.

The goal of this chapter is to justify this intuition. It turns out that the precise mathematical justification is not very simple. In fact, the intuition is not correct for many networks. That is, there are networks for which a natural calculation of the rates of flows indicates that the capacity of no link is exceeded, but the queues in the network grow without bound. We recall such an example below. Thus, it is useful, at least for researchers who need to be precise mathematically, to clarify when the rate conditions suffice to guarantee stability.

In this chapter, we explain these results in a form that we hope is more accessible than the standard treatment. The discussion assumes only an understanding of Markov chains on a countable state space. Nevertheless, the material in the chapter is more technical than in the others.

Consider a network of servers. Each server is equipped with one or more queues. Jobs arrive from outside. Each job joins the queue of one server. After being served, in a first-come, first-served manner by the server, a job may join the queue of another server or possibly leave the system. This routing through the network may be deterministic, i.e., such that each job goes through a specific sequence of servers. It may also be randomized, where a job has some probability of moving from one server to another. It is natural to ask the following question. Given a particular network of servers, under what conditions on the incoming traffic is the network stable? The answer is more complicated than one might think. Intuition suggests that if work arrives for each server at a rate less than the rate of the server, then the system should be stable.

The example of Figure 6.1, borrowed from [Lu and Kumar 1991] shows that this intuition is not correct for general systems. The bottom part of the figure shows a network with four queues.

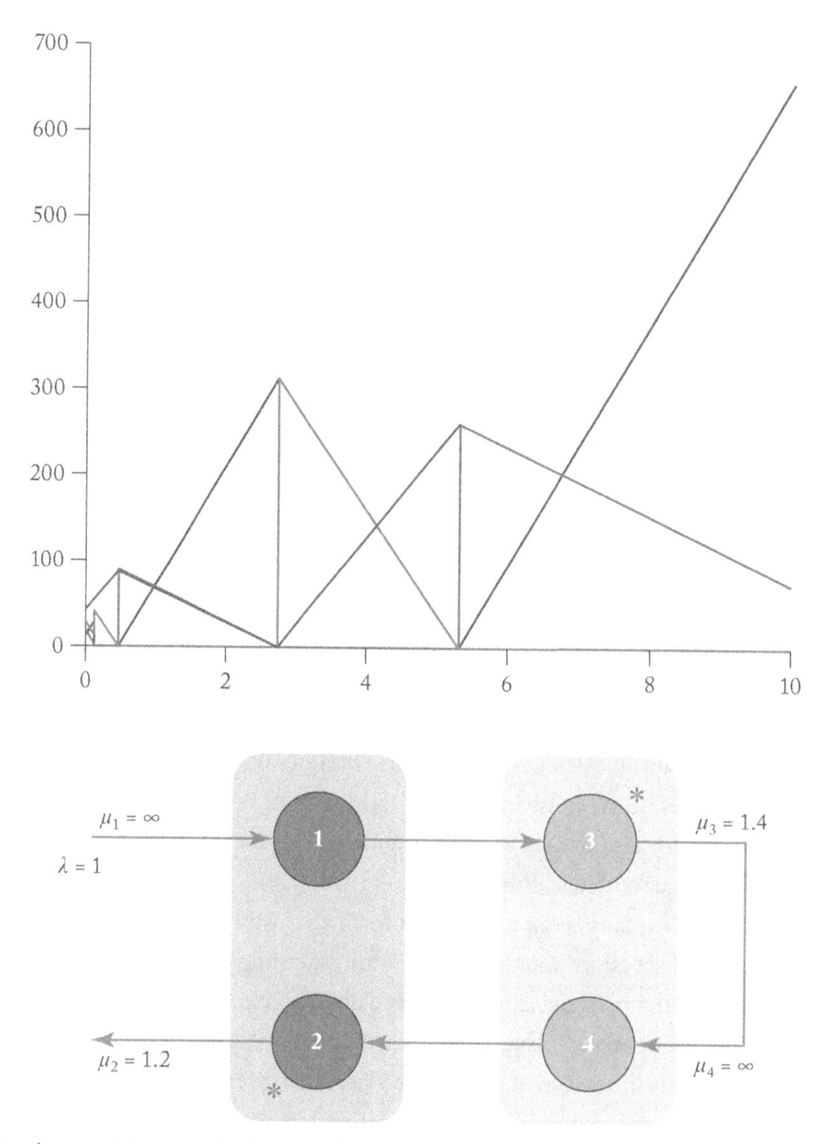

FIGURE 6.1: An unstable network where work arrives for each server at a rate less than the rate of the server.

Queues 1, 2 are served by one server with priority to queue 2 and queues 3, 4 are served by another server with priority to queue 3. The service rates and the arrival rate are shown in the figure. The top part of the figure shows a simulation of the backlog as a function of time and indicates that the network is unstable, even though work arrives at rate less than one for each server.

As another example, we explained in Section 4.4 that Maximum Weighted Matching is stable, but that Maximal Matching may be unstable. Both of these policies are work-conserving (each server in the system works at its maximum rate whenever there are any packets in the queue), but it is still true that one is stable and other isn't. In fact, there are many situations where stable scheduling algorithms are unknown, for instance in data centers. As yet another illustration that stability is not a simple property, [Dimakis and Walrand 2006] provides an example where the system is unstable when the arrivals are deterministic and stable when they are random. This is the type of fragility that one hopes to avoid. See also the excellent discussion of stability in [Bramson 2008].

In this chapter we describe a technique introduced in [Dai 1995] for charactering stability that is based on treating the traffic as a fluid rather than a time ordered sequence of packets. Intuitively, the fine details of what happens to individual packets should not affect the stability of a well-designed system. Indeed, if stability depends on complex features of distributions of random variables, then no sensible engineer should trust the design. One hopes that stability depends only on the average arrival rates of traffic and the average service rates. Thus, like watching freeway traffic from an airplane, one hopes that customers flow smoothly like a fluid through the system and that conditions on average rates suffice to guarantee that no bottleneck gets formed.

More precisely, one hopes that if the system is designed so that average flows or fluids go through the system, then fluctuations around the averages will not build up to create adverse macroscopic consequences. Thus, the goal is to characterize systems in which the random flows track the average fluids and that fluctuations are negligible at the macro scale. (As we know from experience, traffic networks do not have that nice property and they are susceptible to congestion collapse.)

Standard treatments of this material use Markov chains with general state spaces. In this chapter, we explain the theory by using only countable Markov chains.

The queueing systems we analyze are of the following type.

- The system consists of a network of such queues, where the path a given packet takes is determined by randomized routing.
- Each queue has i.i.d. service times and a renewal arrival process (i.e., a GI/GI/1 queue) whose arrival rate is strictly smaller than the service rate.

We will show, by employing a technique that treats the traffic as fluid, that any such network is stable when the interarrival and service times are *phase-type*, and when the natural rate conditions are satisfied. Roughly speaking, a phase-type distribution is one that is constructed by a sequence of exponential distributions. Any probability distribution can be approximated to arbitrary accuracy with phase-type distribution so there is very little lost by making the assumption, and it greatly simplifies the arguments. The general result holds when the random times are *non-arithmetic*, i.e., there is no real number such that interarrival times and service times are integer multiples of that number. However, proving the result under these assumptions is considerably more complicated.

In Section 6.1, we outline the argument in the case of a simple network. The key idea is to consider the *fluid limit* of the network. This limit is obtained by scaling time and space, i.e., by looking at the trajectories from afar (zooming out). One shows that if the fluid limit converges to zero in a time proportional to the initial state, then the Markov chain that models the network must come back to a finite set of states after a finite mean time, which implies its stability. We use a model where the service times and interarrival times are approximated by phase-type distributions. Section 6.2 reviews these distributions. They enable us to consider only Markov chains with a countable state space.

6.1 SKETCH OF ARGUMENT

The key idea is fairly intuitive. We look at the average rates of flows and show that the fluctuations around the averages are negligible in the macroscopic scale. That is, in contrast with freeway traffic or with the Lu-Kumar network, there is no congestion collapse where small fluctuations get amplified by the dynamics and result in a macroscopic bottleneck. Thus, the key idea is to show that the queue lengths track the average behavior that they would have if the random arrivals and service rates were replaced by constant rates. The presentation assumes a basic knowledge of continuous time Markov chains.

To avoid notational clutter, we present the main argument only for a simple example. The same argument holds for a general network.

Model 6.1 Consider the network shown in Figure 6.2. The arrivals A_t and B_t are independent renewal processes. That is, the times between successive arrivals at each queue are independent and identically distributed (i.i.d.) and A_t, B_T are the number of arrivals in $[0, t]$. The service times $\{a_n, n \geq 1\}$ in queue 1 and $\{b_n, n \geq 1\}$ in queue 2 are independent and i.i.d. in each queue. Customers that are served in queue 2 go back to queue 1 with probability p, independently of one another and of the current state of the system, or they leave the network.

Let $\mu_1 = 1/E(a_1)$ and $\mu_2 = 1/E(b_1)$ be the service rates in the two queues and λ_1 be the rate of A_t and λ_2 the rate of B_t. Let γ_1 and γ_2 be the average rate of customers that go through queues 1 and 2, respectively, assuming that the queues are stable. The flow conservation equations are

$$\gamma_1 = \lambda_1 + p\gamma_2 \quad \text{and} \quad \gamma_2 = \lambda_2 + \gamma_1.$$

The solution is

$$\gamma_1 = \frac{\lambda_1 + p\lambda_2}{1 - p} \quad \text{and} \quad \gamma_2 = \frac{\lambda_1 + \lambda_2}{1 - p}.$$

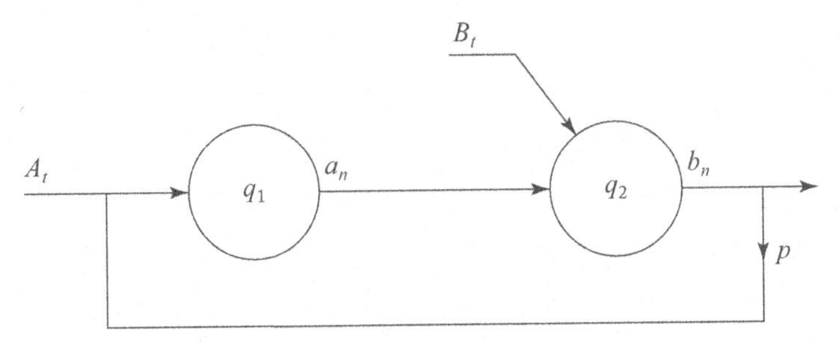

FIGURE 6.2: A representative network with independent renewal arrival processes A_t and B_t, randomized routing, and independent i.i.d. service times a_n and b_n in the two queues.

STABILITY

Assume that

$$\gamma_1 < \mu_1 \quad \text{and} \quad \gamma_2 < \mu_2.$$

Assume also that the interarrival and service times are *phase-type*. We explain that assumption in Section 6.2.

Under these assumptions, the state of the network is an irreducible positive recurrent Markov chain, meaning that it visits every state frequently. In particular, the queues are empty for a positive fraction of the time and the system is asymptotically stationary.

Note that the flow conservation equations assume that the queues are stable since they assume that the rate out of each queue is equal to the rate into the queue. Remarkably, using the solution of the flow conservation equations enables to establish that the queues are stable.

The fundamental reason why this works is that the network belongs to a particular class of queueing systems where the trajectories are continuous functionals of the arrivals and services, continuity being defined in a specific sense. Intuitively, changing the arrivals and the service times slightly leads to a small change in queue lengths: there is no amplification of fluctuations. Thus, the trajectories of the stochastic system are close to those of the fluid limit.

FLUID LIMIT

Let $q_i(t)$ be the queue length of queue i at time t. Also, $q(t) = (q_1(t), q_2(t))$ is the vector of queue lengths at time t. We denote by $||q(t)||$ the Euclidean norm of $q(t)$. The fluid limit is essentially the graph of $q(t)$ as a function of time, after we zoom out by a factor $||q(0)||$ and we let $||q(0)||$ go to infinity. Figure 6.3 shows one realization the queue lengths with arrival rates $\lambda_1 = 3$, $\lambda_2 = 2$, service rates $\mu_1 = 14$, $\mu_2 = 12$ and $p = 0.5$. In the figure, we chose $q_1(0) = 4{,}000$ and $q_2(0) = 3{,}000$, so that $q(0) = bz(0)$ with $b = 5{,}000 = ||q(0)||$ and $z(0) = (0.8, 0.6)$. If we zoom out on the graph by

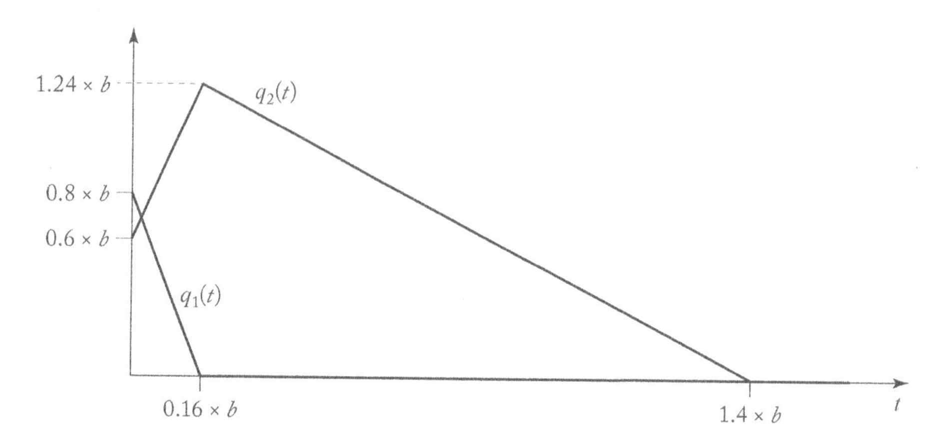

FIGURE 6.3: The queue lengths as a function of time. Here, $b = 5,000$ is very large so that the discreteness of the processes is not visible in the graphs.

scaling the horizontal and vertical axes by a factor b, we are looking at

$$z(t) = \frac{q(bt)}{b}.$$

In this scaling, everything happens 5,000 times faster and the queue lengths are divided by 5,000. The intuition is that, by the strong law of large numbers, the number of arrivals in Nt seconds divided by N should be close to the arrival rate times t, when N is large. The same is also true for services. Thus, in this scaling, a queue changes at a rate equal to the difference between arrival and service rates.

Figure 6.4 provides another view of the scaled vector of queue lengths $z(t)$ as a function of time. More precisely, one should think of $z(t)$ as the limit of $q(bt)/b$ started with $q(0) = bz(0)$, as b goes to infinity.

STABILITY ARGUMENT

We see that the fluid limit $z(t)$ is equal to zero for $t \geq 1.4$. More generally, if we choose any vector $z(0) \geq 0$, the resulting fluid limit $z(t)$ will reach 0 after a finite time equal to $t_0||z(0)||$. Thus, the time to reach 0 is proportional to the norm of the initial state. For the parameters of Figure 6.3, one can check that $t_0 = 1.4$ and this maximum time corresponds to $z(0) = (||z(0)||, 0)$. Thus, $||z(0)||t_0$ is an upper bound on the time for $z(t)$ to hit 0.

As we explain below, the convergence to zero of $z(t)$ guarantees that, for a given $q(0)$, the queue length $q(t)$ of the stochastic system must enter the finite set $\mathcal{S} := \{q \mid ||q|| \leq b\}$, for some b large enough, after a random time τ that has a finite mean value. Every time $q(t)$ leaves the set $\mathcal{S}$,

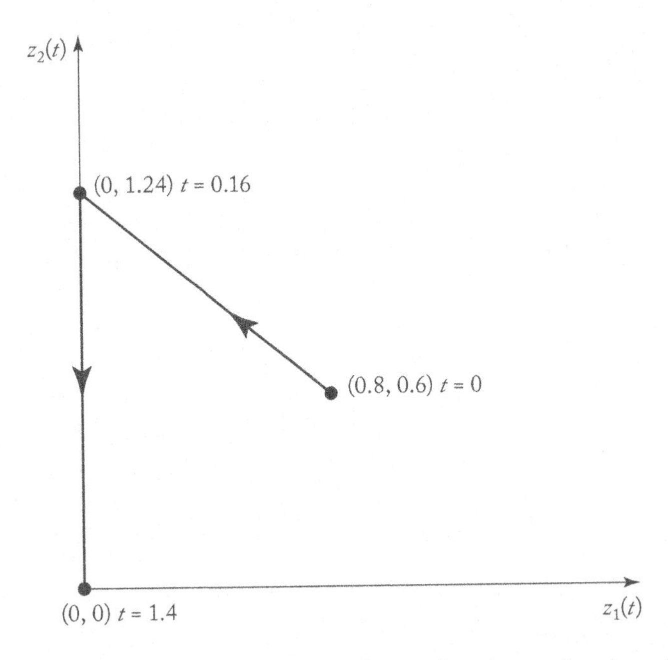

FIGURE 6.4: Another view of the scaled queue lengths as a function of time.

it comes back to it after a finite mean time. As we explain below, this implies that the state of the system spends a positive fraction of time in a finite set of states, so that the Markov chain must be positive recurrent.

The main idea to prove that $q(t)$ comes back to $\mathcal{S}$ is simple. Pick $z(0)$ with $||z(0)|| = 1$. For any given b, we construct the queue lengths starting with $q(0) = bz(0)$ and we note that

$$q(bt)/b \approx z(t) \approx 0, \quad \text{if } b \gg 1 \text{ and } t \geq t_0. \tag{6.1}$$

Figure 6.3 illustrates this fact, with $t_0 = 1.4$.

One can show that the previous result also holds in expectation, i.e.,

$$E(||q(bt)/b||) \approx 0, \quad \text{if } b \gg 1 \text{ and } t \geq t_0. \tag{6.2}$$

This fact is proved formally at the end of the chapter.

This shows that there is some b_0 large enough so that

$$E(||q(t)||) \leq (1 - \epsilon)||q(0)||, \quad \text{if } ||q(0)|| \geq b_0 \text{ and } t \geq ||q(0)||t_0. \tag{6.3}$$

Thus, if the initial queue lengths are large enough, they shrink (in expected norm) after a time proportional to the initial norm.

Let then $t_1 = t(q(0)) = ||q(0)||t_0$. We see that, if $||q(0)|| > b_0$,

$$E(||q(t_1)||) \leq ||q(0)|| - \beta t(q(0)),$$

where $\beta = \epsilon/t_0$.

Assume then that $||q(t)|| > b_0$ at times $t \in \{0, t_1 = t(q(0)), t_2 = t(q(t_1)), \ldots, t_n = t(q(t_{n-1}))\}$. Let us call this event $\mathcal{F}$. Then, using the inequality above repeatedly, we see that

$$E(||q(t_n)||1_{\mathcal{F}}) \leq ||q(0)|| - \beta E(t_n 1_{\mathcal{F}}).$$

Now, say that τ is the first value of t_{n+1} such that $||q(t_{n+1})|| \leq b_0$ and let $\tau = \infty$ if there is no such t_{n+1}. We then conclude that

$$E(||q(\tau \wedge n)||) \leq ||q(0)|| - \beta E(\tau \wedge n), \forall n \geq 1.$$

Since the left-hand side is positive, we see that $E(\tau \wedge n) \leq ||q(0)||/\beta$. Letting $n \to \infty$, we conclude, using the Monotone Convergence Theorem, that $E(\tau)$ is finite.

We provide some technical details in the next section for the interested readers.

6.2 PROOF OF STABILITY

The contribution of this section is to prove the stability when the random times are phase-type. This simplifies the analysis because the state of the system is then a Markov chain with a countable state space. Stability is the usual positive recurrence. The key idea is to use the fluid limit to prove stability of the queueing system.

The proof follows the general argument that we outlined in the last section. We first clarify what is meant by phase-type distributions.

PHASE-TYPE

Before defining phase-type distributions, we look at a few examples. Consider a queue, shown in Figure 6.5, where the service times are independent and uniformly distributed in [0, 1] and the times between successive arrivals are independent and uniformly distributed in [0, 1.2]. For the purpose of analysis, we would like to describe this system by a Markov chain. The state of the system at time t must specifies how many jobs there are in the queue and also how long the job currently served has been in service and how long it has been since the last arrival. This information is needed to predict the evolution of the system after time t. Thus, the set of possible values of the state is not countable.

Let us approximate a service time as follows (see Figure 6.6). There are $N \gg 1$ *phases* of service. Each phase has an exponential duration with rate N. When a job starts service, it starts in a phase i

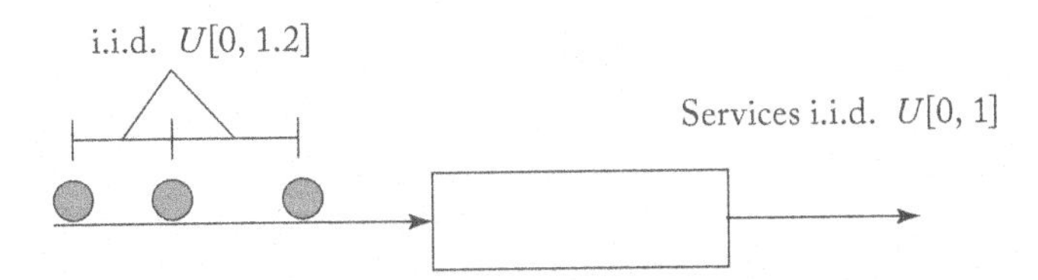

FIGURE 6.5: A queue with uniform interarrival and service times.

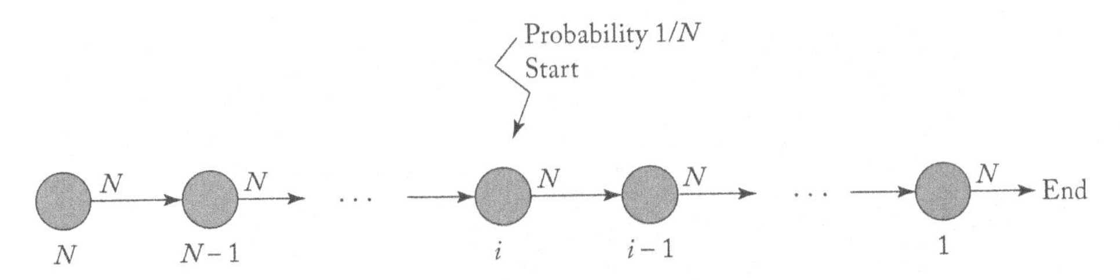

FIGURE 6.6: Phase-type approximation of uniform service time.

that is chosen uniformly in $\{1, \ldots, N\}$. The job then goes through the phases $\{i, i-1, i-2, \ldots, 1\}$ and the service completes when phase 1 is done. The state of the service is now specified by the value of i, which is the remaining number of phases of service that the job must go through. With probability p, the job starts in one of the phases $\{1, 2, \ldots, pN\}$ and goes through at most pN phases of service that are exponentially distributed with rate N. Hence,with probability p, the service time is less than the sum of pN i.i.d. random variables that are exponentially distributed with rate N. These random variables can be written as $X_1/N, X_2/N, \ldots, X_{pN}/N$ where the X_m are i.i.d. and exponentially distributed with rate 1. By the (weak) law of large numbers, this sum of random variables is almost equal to p. Thus, with probability p, the service time is less than p, so that the service time is almost uniformly distributed in $[0, 1]$.

We have replaced a uniform random variable by a random number of exponential phases. The benefit is that we have a finite description of the state of service. We can do the same with the interarrival times. We now have a description $x_t = (\phi_t, \xi_t, q_t)$ of the queue, where $\phi_t \in \{1, \ldots, N\}$ is the phase of service of the job being served (0 if there is no job in the queue), $\xi_t \in \{1, \ldots, M\}$ is the phase of the interarrival time, and $q_t \in \{0, 1, \ldots\}$ is the number of jobs in the queue. Note that x_t is now a Markov chain on a countable state space. Our goal was certainly not to get a compact representation of the state. Instead, it is to enable us to use results from countable Markov chains to study the stability of the system.

Distributions of random times can be approximated as closely as desired by a similar procedure. Figure 6.7 shows a representative situation. The top of the figure shows that a job starts in phase 1,

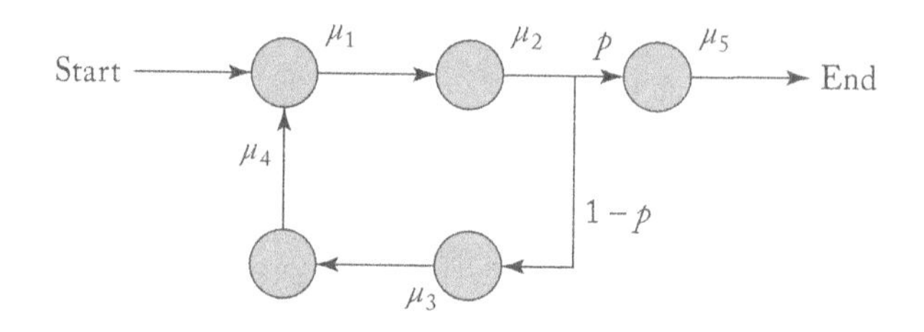

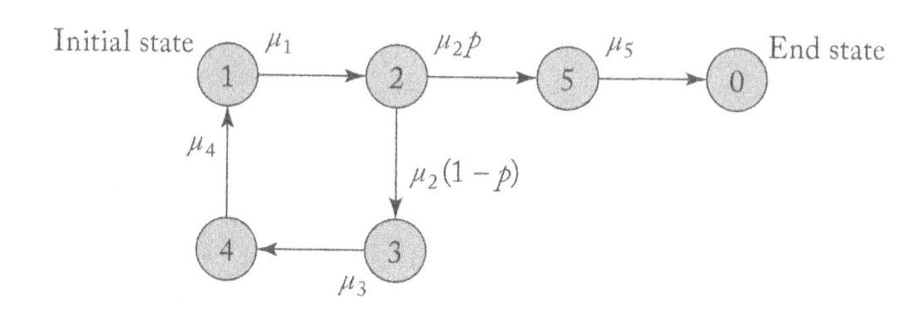

FIGURE 6.7: Phases of service (top) and Markov chain representation (bottom).

which has an exponential distribution with rate μ_1, then moves to phase 2 that has an exponential distribution with rate μ_2. When phase 2 is completed, with probability p, the job moves to phase 5. With probability $1 - p$, the job moves to phase 3, and so on. The bottom part of the figure shows that the service time of this job can be viewed as the time when the Markov chain with states $\{0, 1, \ldots, 5\}$, started in state 1, reaches state 0.

More generally, consider a finite irreducible Markov chain $\{\xi(t), t \geq 0\}$ on $\{0, 1, \ldots, N\}$ with rate matrix Q and initial distribution ϕ on $\{1, \ldots, N\}$. Then, the first passage time τ to state 0, i.e.,

$$\tau = \inf\{t \geq 0 \mid \xi(t) = 0\}$$

is said to have a phase-type distribution.

With a large enough number of states, the phase type distribution can approximate any practical distribution of a random time. A systematic method is to consider the Laplace transform of the distribution of the random time and to approximate it by a rational function. The rational function can then be mapped into a phase-type distribution. The mechanics of this approximation do not matter for our discussion. Suffices it to say that there is no loss of practical relevance by assuming that the interarrival and service times of the network have a phase type distribution. We make that assumption in this chapter so that we can model the networks by countable Markov chains.

At any given time t, the state $x(t)$ of the network can be written as $x(t) = (q(t), y(t))$ where $q(t)$ is the vector of queue lengths and $y(t)$ specifies the phases of the interarrival times and service times. Thus, $x(t)$ is a countable Markov chain. Also, one can verify that this Markov chain is irreducible. This is very convenient because irreducible Markov chains are very simple: either they spend a positive fraction of time in any finite set of states or they spend a zero fraction of time in any finite set of states. In the first case, the Markov chain is positive recurrent, it has a unique invariant distribution and it is asymptotically stationary. In this case, the network is stable. Thus, to prove the stability of the network, it suffices to show that there is some finite set of states where it spends a positive fraction of time. For us, that set is the collection of states $x = (q, y)$ where $||q|| \leq b_0$ for some b_0. There are finitely many such states because y takes finitely many values.

Note a useful property of renewal processes with a phase-type distribution of interarrival times: they have a bounded rate. For instance, the rate of A_t is bounded by the maximum rate of transition to state 0 of its phase process. Consequently, we see that $A_t + B_t$ is smaller than some Poisson process R_t with rate ρ.

In particular, we see that

$$q_1(t) + q_2(t) \leq q_1(0) + q_2(0) + A_t + B_t \leq q_1(0) + q_2(0) + R_t, \forall t \geq 0.$$

This implies that, with $k = q_1(0) + q_2(0)$,

$$E(||q(t)||) \leq E(q_1(t) + q_2(t)) \leq k + \rho t, \tag{6.4}$$

which says that the norm of the queue length vector grows at most linearly over time.

Also,

$$\begin{aligned} E(||q(t)||^2) &\leq (q_1(t) + q_2(t))^2 \leq k^2 + 2kE(R_t) + E(R_t^2) \\ &\leq k^2 + 2k\rho t + ((\rho t)^2 + \rho t) = k^2 + (2k+1)\rho t + \rho^2 t^2. \end{aligned}$$

Consequently, for any fixed $t > 0$, with $V_b = ||q(bt)/b||$, one sees that

$$E(V_b^2) \leq \frac{k^2 + (2k+1)\rho b t + \rho^2 b^2 t^2}{b^2} < \kappa, \forall b,$$

for some constant κ that depends on t. Thus, the mean value of the square of the random variables V_b are bounded. This implies that they are uniformly integrable. As a consequence, if we can show that $V_b \to 0$ almost surely, we conclude that $E(V_b) \to 0$. We then use the fact that if $t > ||q(0)||t_0$, then $V_b \to 0$ almost surely. This fact then proves (6.2).

Proof of Theorem 6.1. The convergence of the scaled queue lengths to the fluid limit follows from the corresponding result for renewal processes and from the fact that the scaled queue lengths converge to a scaled continuous functional of the arrival and service processes.

Let A_t be a renewal process with rate λ. For $b > 0$, let $a_t^b = A_{bt}/b$. Then, for all $T > 0$,

$$\sup\{|a_t^b - \lambda t|, 0 \leq t \leq T\} \to 0, \quad \text{almost surely as } b \to \infty.$$

Thus, the scaled renewal process converges to a straight line. The convergence is uniform on finite sets $[0, T]$ and almost sure. This result follows from the fact that the finite dimensional distributions of the scaled process converge to those of the straight line, by the strong law of large numbers, and from the fact that the scaled renewal processes are *tight*. For details, see [Billingsley 1999].

Let S_t^1 and S_t^2 be the renewal processes with interarrival times a_n and b_n, respectively. One can construct fictitious queue lengths $h(t)$ as a function of these processes as follows. Pretend that queue i completes a service whenever the process S_t^i jumps and there is a customer in queue i. The resulting queue lengths $h(t)$ are not exactly equal to $q(t)$ because in $q(t)$, a service starts only when a customer is in the queue whereas in the construction of $h(t)$ it starts before a customer is in the queue. However, one can show that $h(bt)/b \approx q(bt)/b$. Also, $h(t)$ is a continuous functional of the renewal processes A, B, S^1, and S^2. ■

6.3 SUMMARY AND REFERENCES

This chapter is somewhat different in style from the others, so we feel that we should explain why we included it in this text.

In our study of distributed protocols for sharing the links of a network, we focused on deterministic models where the flows are characterized by a rate, say x bits per second. We explained how to select the rates of these flows so that they meet the capacity constraints of the links. Intuitively, this condition means that the network is "stable." However, we know that the actual flows of packets in a network are variable. A video stream has a variable rate; web transfers are bursty, and so on.

The connection between the stability of the system with variable flows and the conditions on the average rate of fluids is made precise by the results of this chapter. This relationship was demonstrated by Jim Dai [Dai 1995] in great generality. That theory is rather technical and understanding it requires studying Harris recurrent Markov chains on general state spaces. A good presentation of that theory can be found in Meyn and Tweedie [Meyn and Tweedie 2009]. Our objective was to make the theory accessible to readers who are familiar only with Markov chains with a countable state space.

Bibliography

[1] A. Archer and É. Tardos. Frugal path mechanisms. *ACM Transactions on Algorithms (TALG)*, 3(1):3, 2007. DOI: 10.1145/1186810.1186813. 57

[2] K. J. Arrow. A difficulty in the concept of social welfare. *Journal of Political Economy*, 58(4):328–346, 1950. DOI: 10.1086/256963. 3

[3] L. M. Ausubel. An efficient ascending-bid auction for multiple objects. *American Economic Review*, 94(5):1452–1475, 2004. DOI: 10.1257/0002828043052330. 59

[4] D. P. Bertsekas and R. Gallager. *Data networks*. Prentice-Hall International Englewood Cliffs, NJ, 1987. xvii

[5] P. Billingsley. *Convergence of Probability Measures, 2nd Edition*. Wiley, 1999. DOI: 10.1002/9780470316962. 122

[6] K. Binmore. Interpersonal Comparison of Utility. *Oxford handbook of the philosophy of economic science*, pages 540–559, 2009. DOI: 10.1.1.177.5114. 17

[7] G. Birkhoff. Three observations on linear algebra. *Univ. Nac. Tacuman Rev. Ser. A*, 5:147–151, 1946. 88

[8] P. Bolton and M. Dewatripont. *Contract theory*. MIT Press: Cambridge and London, 2005. 110

[9] S. P. Boyd and L. Vandenberghe. *Convex optimization*. Cambridge University Press, 2004. 23

[10] M. Bramson. *Stability of Queueing Networks*, volume 5. Probability Surveys, 2008. DOI: 10.1214/08-PS137. 113

[11] S. T. Chuang, A. Goel, N. McKeown, and B. Prabhakar. Matching output queueing with a combined input/output-queued switch. *IEEE Journal on Selected Areas in Communications*, 17(6):1030–1039, 1999. DOI: 10.1109/49.772430. 83, 85

[12] E. Clarke. Multipart pricing of public goods. *Public Choice*, 11(1):17–33, 1971. DOI: 10.1007/BF01726210. 71

[13] P. Cramton and J. A. Schwartz. Collusive bidding in the fcc spectrum auctions. *The BE Journal of Economic Analysis & Policy*, 1(1):11, 2002. 67

[14] J. Dai. On positive Harris recurrence of multiclass queueing networks: a unified approach via fluid limit models. *Annals of Applied Probability*, 5:49–77. 1995. 113, 122

[15] P. Diaconis. A generalization of spectral analysis with application to ranked data. *The Annals of Statistics*, 17(3):949–979, 1989. DOI: 10.1214/aos/1176347251. 8

[16] A. Dimakis and J. Walrand. Sufficient conditions for stability of longest-queue-first scheduling: second-order properties using fluid limits. *Advances in Applied Probability*, 38:505–521, 2006. DOI: 10.1239/aap/1151337082. 113

[17] L. Duan, T. Kubo, K. Sugiyama, J. Huang, T. Hasegawa, and J. Walrand. Motivating smartphone users' collaborations in data acquisitions and distributed computing. Under review. *IEEE Transactions on Mobile Computing*, 2013. 108

[18] B. Edelman, M. Ostrovsky, M. Schwarz, T. D. Fudenberg, L. Kaplow, R. Lee, P. Milgrom, M. Niederle, and A. Pakes. Internet advertising and the generalized second price auction: Selling billions of dollars worth of keywords. *American Economic Review*, 97, 2005. 71

[19] J. Elster and J. E. Roemer. *Interpersonal comparisons of well-being*. Cambridge University Press, 1991. DOI: 10.1017/CBO9781139172387. 17

[20] A. Eryilmaz, R. Srikant, and J. Perkins. Stable scheduling policies for fading wireless channels. *IEEE/ACM Transactions on Networking*, pages 411–424, 2005. DOI: 10.1109/TNET.2004.842226. 88

[21] R. Etkin, A. Parekh, and D. Tse. Spectrum sharing for unlicensed bands. *IEEE Journal on selected areas in communications*, 25(3):517–528, 2007. DOI: 10.1109/JSAC.2007.070402. 36, 48, 49, 51

[22] P. C. Fishburn. Nontransitive preferences in decision theory. *Journal of Risk and Uncertainty*, 4(2):113–134, 1991. DOI: 10.1007/BF00056121. 3

[23] J. W. Friedman. A non-cooperative equilibrium for supergames. *The Review of Economic Studies*, 38(1):1–12, 1971. DOI: 10.2307/2296617. 49

[24] J. W. Friedman. *Oligopoly and the Theory of Games*. North-Holland Publishing Company, 1977. 49

[25] V. Gajic, J. Huang, and B. Rimoldi. Competition of wireless providers for atomic users: Equilibrium and social optimality. In *Communication, Control, and Computing, 2009. Allerton 2009. 47th Annual Allerton Conference on*, pages 1203–1210. IEEE, 2009. 27

[26] D. Gale and L. S. Shapley. College admissions and the stability of marriage. *American Mathematical Monthly*, 69(1):9–15, 1962. DOI: 10.2307/2312726. 79, 94

[27] L. Georgiadis, R. Guerin, and A. Parekh. Optimal multiplexing on a single link: delay and buffer requirements. *IEEE Transactions on Information Theory*, 43(5):1518–1535, 1997. DOI: 10.1109/18.623149. 83

[28] J. Green and J. J. Laffont. Characterization of satisfactory mechanisms for the revelation of preferences for public goods. *Econometrica*, 45(2):427–438, 1977. DOI: 10.2307/1911219. 58

[29] T. Groves. Incentives in teams. *Econometrica*, 41(4):617–631, 1975. DOI: 10.2307/1914085. 71

[30] J. C. Harsanyi. Interpersonal utility comparisons. *The New Palgrave: utility and probability*, page 128, 1990. 17

[31] S. Hart and P. J. Reny. Maximal revenue with multiple goods: Nonmonotonicity and other observations. *The Hebrew University of Jerusalem, Center for Rationality*, DP-630, 2012. 62

[32] C. Hauert. Public Good Games. *Virtual Labs*, (2.1), 2005. 96

[33] L. Huang and J. C. Walrand. A benes packet network. In *INFOCOM*, pages 1204–1212, 2013. 51

[34] S. Jagabathula and D. Shah. Inferring rankings under constrained sensing. In *Advances in Neural Information Processing Systems*, pages 753–760, 2008. DOI: 10.1.1.143.6853. 8

[35] B. Ji, C. Joo, and N. B. Shroff. Delay-based back-pressure scheduling in multihop wireless networks, to appear. *IEEE/ACM Trans. on Networking*, 2013. DOI: 10.1109/TNET.2012.2227790. 88

[36] L. Jiang and J. Walrand. *Scheduling and Congestion Control for Wireless and Processing Networks*. Morgan-Claypool, 2010. 40, 51, 89, 94

[37] C. Joe-Wong, S. Sen, T. Lan, and M. Chiang. Multi-resource allocation: Fairness-efficiency tradeoffs in a unifying framework. In *INFOCOM, 2012 Proceedings IEEE*, pages 1206–1214. IEEE, 2012. 36

[38] R. Johari and J. N. Tsitsiklis. Efficiency loss in a network resource allocation game. *Mathematics of Operations Research*, pages 407–435, 2004. DOI: 10.1287/moor.1040.0091. 43, 51

[39] R. Johari and J. N. Tsitsiklis. Efficiency of scalar-parameterized mechanisms. *Operations Research*, pages 823–839, 2009. DOI: 10.1287/opre.1080.0638. 42

[40] F. P. Kelly. Charging and rate control for elastic traffic. *European Transactions on Telecommunications*, 8:33–37, 1997. DOI: 10.1002/ett.4460080106. 17, 51

[41] L. Kleinrock. Queueing systems, volume I: theory, 1975. DOI: 10.1109/TCOM.1977.1093722. 19

[42] P. Klemperer. Auction theory: A guide to the literature. *Journal of Economic Surveys*, 13(3):227–286, 1999. DOI: 10.1111/1467-6419.00083. 65

[43] D. M. Kreps. *Notes on the Theory of Choice*, volume 2. Westview press Boulder, 1988. 17

[44] V. Krishna. *Auction Theory, Second Edition*. Academic Press, 2009. 71

[45] J. F. Kurose, K. W. Ross, and K. Ross. *Computer networking: a top-down approach featuring the Internet*. Addison-Wesley Reading, MA, 2003. xvii

[46] T. Lan, D. Kao, M. Chiang, and A. Sabharwal. *An axiomatic theory of fairness in network resource allocation*. IEEE, 2010. DOI: 10.1109/INFCOM.2010.5461911. 36

[47] S. Lu and P. R. Kumar. Distributed scheduling based on due dates and buffer priorities. *IEEE Transactions on Automatic Control*, 36(12):1406–1416, 1991. DOI: 10.1109/9.106156. 111

[48] A. Mas-Colell, M. D. Whinston, and J. R. Green. *Microeconomic theory*. Oxford university press New York, 1995. 4, 17, 58

[49] E. Maskin and J. Riley. Optimal auctions with risk averse buyers. *Econometrica: Journal of the Econometric Society*, 52(6):1473–1518, 1984. DOI: 10.2307/1913516. 64

[50] E. S. Maskin and J. G. Riley. Auction theory with private values. *The American Economic Review*, 75(2):150–155, 1985. 63

[51] A. Maulloo, F. P. Kelly, and D. Tan. Rate control in communication networks: shadow prices, proportional fairness and stability. *Journal of the Operational Research Society*, 49:237–252, 1998. 17

[52] R. P. McAfee and J. McMillan. Auctions and bidding. *Journal of economic literature*, 25(2):699–738, 1987. 64, 65

[53] N. McKeown, P. Varaiya, and J. Walrand. Scheduling cells in an input-queued switch. *IEEE Electronics Letters*, Series A:2174–5, 1993. DOI: 10.1049/el:19931459. 89

[54] N. McKeown, V. Anantharam, and J. Walrand. Achieving 100 throughput in an input queued switch. In *IEEE INFOCOM*, pages 296–302, 1996. DOI: 10.1109/INFCOM.1996.497906. 86, 94

[55] R. Metcalfe. Packet Communication. *MIT Project MAC Technical Report MAC TR-114*, 1973. 37

[56] S. Meyn and R. Tweedie. *Markov Chains and Stochastic Stability, Second Edition*. Cambridge Mathematical Library, 2009. DOI: 10.1017/CBO9780511626630. 122

[57] P. Milgrom. Putting auction theory to work: The simultaneous ascending auction. In *Journal of Political Economy*. Citeseer, 1999. 66

[58] J. Mo and J. Walrand. Fair end-to-end window-based congestion control. *IEEE/ACM Transactions on Networking (ToN)*, 8(5):556–567, 2000. DOI: 10.1109/90.879343. 25, 51

[59] H. Moulin and S. Shenker. Strategyproof sharing of submodular costs: budget balance versus efficiency. *Economic Theory*, 18(3):511–533, 2001. DOI: 10.1007/PL00004200. 97, 99, 104

[60] R. B. Myerson. Optimal auction design. *Mathematics of operations research*, 6(1):58–73, 1981. DOI: 10.1287/moor.6.1.58. 60, 61, 64, 67, 71

[61] J. Nash. The bargaining problem. *Econometrica*, 18(2):155–162, 1950. DOI: 10.2307/1907266. 13, 17, 96, 110

[62] M. Neely. *Stochastic Network Optimization with Application to Communication and Queueing Systems*. Morgan-Claypool, 2010. 51

[63] M. J. Neely, E. Modiano, and C. P. Li. Fairness and Optimal Stochastic Control for Heterogeneous Networks. *Proceedings of IEEE Infocom*, 2005. 51

[64] A. K. Parekh and R. G. Gallager. A generalized processor sharing approach to flow control inintegrated services networks: the multiple node case. *IEEE/ACM transactions on networking*, 2(2):137–150, 1994. DOI: 10.1109/90.298432. 19

[65] D. Parfit. *Reasons and persons*. Oxford University Press, USA, 1986. DOI: 10.1093/019824908X.001.0001. 12

[66] B. Peleg and P. Sudhölter. *Introduction to the theory of cooperative games*. Springer Verlag, 2007. 13

[67] L. L. Peterson and B. S. Davie. *Computer networks: a systems approach*. Morgan Kaufmann, 2007. xvii

[68] M. D. Plummer and L. Lovász. *Matching theory*. Access Online via Elsevier, 1986. 73

[69] W. Poundstone. *Gaming the Vote: Why Elections Aren't Fair (and what We Can Do about It)*. Hill & Wang, 2008. 5

[70] L. P. Qian, Y. J. A. Zhang, and J. Huang. Mapel: Achieving global optimality for a non-convex wireless power control problem. *Wireless Communications, IEEE Transactions on*, 8(3):1553–1563, 2009. 36

[71] R. W. Rosenthal. Games of perfect information, predatory pricing and the chain-store paradox. *Journal of Economic Theory*, 25:92–100, 1981. DOI: 10.1016/0022-0531(81)90018-1. 107

[72] A. E. Roth. The economics of matching: Stability and incentives. *Mathematics of Operations Research*, 7(4):617–628, 1982. DOI: 10.1287/moor.7.4.617. 80

[73] A. E. Roth. Misrepresentation and stability in the marriage problem. *Journal of Economic Theory*, 34(2):383–387, 1984. DOI: 10.1016/0022-0531(84)90152-2. 80, 94

[74] A. E. Roth. *The Shapley value: essays in honor of Lloyd S. Shapley, Introduction*. Cambridge Univ Pr, 1988. DOI: 10.1017/CBO9780511528446. 94, 103, 110

[75] A. E. Roth, T. Sonmez, and M. U. Unver. Kidney exchange. Working Paper 10002, National Bureau of Economic Research, September 2003. URL http://www.nber.org/papers/w10002. 94

[76] A. Rubinstein. Perfect equilibrium in a bargaining model. *Econometrica*, 50(1):97–109, 1982. DOI: 10.2307/1912531. 96, 110

[77] S. Shakkottai and R. Srikant. *Network optimization and control*. Now Publishers, 2008. 12, 51

[78] L. S. Shapley. Utility comparison and the theory of games. *The Shapley value: essays in honor of Lloyd S. Shapley*, page 307, 1988. 13, 95

[79] S. Shenker. Fundamental design issues for the future Internet. *Selected Areas in Communications, IEEE Journal on*, 13(7):1176–1188, 1995. DOI: 10.1109/49.414637. 17

[80] R. Srikant and L. Ying. *Communication Networks*. Cambridge Univ Pr, 2014. 19, 51

[81] L. Tassiulas and A. Ephremides. Stability properties of constrained queueing systems and scheduling policies for maximum throughput in multihop radio networks. *IEEE Transactions on Automatic Control*, 37(37):1936–1948, 1992. DOI: 10.1109/9.182479. 51, 87, 94

[82] D. Tse and P. Viswanath. *Fundamentals of wireless communication*. Cambridge Univ Pr, 2005. DOI: 10.1017/CBO9780511807213. 34

[83] W. Vickrey. Counterspeculation, auctions, and competitive sealed tenders. *Journal of finance*, pages 8–37, 1961. DOI: 10.1111/j.1540-6261.1961.tb02789.x. 54, 71

[84] J. Walrand and S. Parekh. *Communication Networks: A Concise Introduction*. Morgan-Claypool, 2010. xvii

[85] J. Walrand. *An Introduction to Queueing Networks*. Prentice Hall Englewood Cliffs, NJ, 1988. 19

[86] S. Yang and B. Hajek. VCG-Kelly mechanisms for allocation of divisible goods: Adapting VCG mechanisms to one-dimensional signals. *IEEE Journal on Selected Areas in Communications*, 25:1237–1243, 200. DOI: 10.1109/CISS.2006.286682. 51

Authors' Biographies

ABHAY PAREKH

Abhay Parekh received his Ph.D. in EECS from MIT in 1992. His graduate work was in the area of network resource allocation and papers from his dissertation won the IEEE William Bennett award and the Infocom Best Paper award. After spending time in research at Bell Labs, IBM Thomas J. Watson Research Center, and Sun Microsystems, he cofounded FastForward Networks, which built the first products for application-level multicasting. He has also been on the faculty at the University of California, Berkeley, as an Adjunct Professor of EECS since 2003, where he has worked in areas such as wireless interference management and peer-to-peer networks. In addition to his academic affiliation, Abhay has been a Venture Partner at Accel Partners and the founding CEO of several startup companies.

JEAN WALRAND

Jean Walrand received his Ph.D. in EECS from the University of California, Berkeley, and has been on the faculty of that department since 1982. He is the author of *An Introduction to Queueing Networks* (Prentice Hall, 1988) and *Communication Networks: A First Course* (2nd ed., McGraw-Hill, 1998), and co-author of *High-Performance Communication Networks* (2nd ed., Morgan Kaufman, 2000), *Communication Networks: A Concise Introduction* (Morgan & Claypool, 2010), *Scheduling and Congestion Control for Communication and Processing Networks* (Morgan & Claypool, 2010), and *Probability in Electrical Engineering and Computer Science* (Amazon, 2014). Jean is a Fellow of the Belgian–American Educational Foundation and of the IEEE, and a recipient of the Lanchester Prize, the IEEE Kobayashi Award, and the ACM SIGMETRICS Achievement Award.

Index

Printed in the United States
by Baker & Taylor Publisher Services